Del Vecchio

Understanding Design of Experiments: A Primer for Technologists

D0424486

Hanser **Understanding** Books

A Series of Mini-Tutorials

Series Editor: E.H. Immergut

R.J. Del Vecchio

Understanding
Design of Experiments:
A Primer for Technologists

1999

Hanser Publishers, Munich

Hanser/Gardner Publications, Inc., Cincinnati

The Author:
R.J. Del Vecchio, Technical Consulting Services, 3 John Matthews Road, Southborough, MA
01772, U.S.A.

Distributed in the U.S.A. and in Canada by
Hanser/Gardner Publications, Inc.
6915 Valley Avenue, Cincinnati, Ohio 45244-3029, U.S.A.
Fax: (513) 527-8950
Phone: (513) 527-8977 or 1-800-950-8977
Internet: http://www.hansergardner.com

Distributed in all other countries by
Carl Hanser Verlag
Postfach 86 04 20, 81631 München, Germany
Fax: +49 (89) 98 12 64

The use of general descriptive names, trademarks, etc., in this publication, even if the former are not especially identified, is not to be taken as a sign that such names, as understood by the Trade Marks and Merchandise Marks Act, may accordingly be used freely by anyone.

While the advice and information in this book are believed to be true and accurate at the date of going to press, neither the authors nor the editors nor the publisher can accept any legal responsibility for any errors or omissions that may be made. The publisher makes no warranty, express or implied, with respect to the material contained herein.

Library of Congress Cataloging-in-Publication Data
Del Vecchio, R.J.
Understanding design of experiments / R.J. Del Vecchio
    p.   cm. -- (Hanser understanding books)
Includes bibliographical references and index.
ISBN 1-56990-222-4
1. Experimental design.   I. Title.   II. Series.
QA279.D45   1997
001-4'34--dc21

Die Deutsche Bibliothek - CIP - Einheitsaufnahme
Del Vecchio, R.J.:
Understanding design of experiments / R.J. Del Vecchio. - Munich ;
Vienna ; New York : Hanser ; Cincinnati : Hanser/Gardner, 1977
    ISBN 3-446-18657-3

© Carl Hanser Verlag, Munich 1997
Typeset in the U.S.A. by Marcia D. Sanders, Frederick, MD
Printed and bound in Germany by Druckhaus "Thomas Müntzer", Bad Langensalza

Introduction to the Series

In order to keep up in today's world of rapidly changing technology we need to open our eyes and ears and, most importantly, our minds to new scientific ideas and methods, new engineering approaches and manufacturing technologies and new product design and applications. As students graduate from college and either pursue academic polymer research or start their careers in the plastics industry, they are exposed to problems, materials, instruments and machines that are unfamiliar to them. Similarly, many working scientists and engineers who change jobs must quickly get up to speed in their new environment.

To satisfy the needs of these "newcomers" to various fields of polymer science and plastics engineering, we have invited a number of scientists and engineers, who are experts in their field and also good communicators, to write short, introductory books which let the reader "understand" the topic rather than to overwhelm him/her with a mass of facts and data. We have encouraged our authors to write the kind of book that can be read profitably by a beginner, such as a new company employee or a student, but also by someone familiar with the subject, who will gain new insights and a new perspective,

Over the years this series of **Understanding** books will provide a library of mini-tutorials on a variety of fundamental as well as technical subjects. Each book will serve as a rapid entry point or "short course" to a particular subject and we sincerely hope that the readers will reap immediate benefits when applying this knowledge to their research or work-related problems.

E.H. Immergut
Series Editor

SPE Books from Hanser Publishers

Belofsky, Plastics: Product Design and Process Engineering
Bernhardt, Computer Aided Engineering for Injection Molding
Brostow/Corneliussen, Failure of Plastics
Chan, Polymer Surface Modification and Characterization
Charrier, Polymeric Materials and Processing—Plastics, Elastomers and Composites
Del Vecchio, Understanding Design of Experiments: A Primer for Technologists
Ehrig, Plastics Recycling
Ezrin, Plastics Failure Guide
Gordon, Total Quality Process Control for Injection Molding
Gordon/Shaw, Computer Programs for Rheologists
Gruenwald, Plastics: How Structure Determines Properties
Macosko, Fundamentals of Reaction Injection Molding
Manzione, Applications of Computer Aided Engineering in Injection Molding
Matsuoka, Relaxation Phenomena in Polymers
Menges/Mohren, How to Make Injection Molds
Michaeli, Extrusion Dies for Plastics and Rubber
Michaeli/Greif/Kaufmann/Vossebürger, Training in Plastics Technology
Michaeli/Greif/Kretzschmar/Kaufmann/Bertuleit, Training in Injection Molding
Neuman, Experimental Strategies for Polymer Scientists and Plastics Engineers
O'Brien, Applications of Computer Modeling for Extrusion and Other Continuous
    Polymer Processes
Progelhof/Throne, Polymer Engineering Principles
Rauwendaal, Polymer Extrusion
Rees, Mold Engineering
Rosato, Designing with Reinforced Composites
Saechtling, International Plastics Handbook for the Technologist, Engineer and User
Stevenson, Innovation in Polymer Processing
Stoeckhert, Mold-Making Handbook for the Plastics Engineer
Throne, Thermoforming
Tucker, Fundamentals of Computer Modeling for Polymer Processing
Ulrich, Introduction to Industrial Polymers
Wright, Injection/Transfer Molding of Thermosetting Plastics
Wright, Molded Thermosets: A Handbook for Plastics Engineers, Molders and
    Designers

Foreword

The Society of Plastics Engineers is pleased to sponsor and endorse *Design of Experiments: A Primer for Technologists* by R.J. Del Vecchio. This presentation provides a clearly written, uncomplicated, yet thorough understanding necessary for implementing design experiments by working technologists.

SPE, through its Technical Volumes Committee, has long sponsored books on various aspects of plastics. Its involvement has ranged from identification of needed volumes and recruitment of authors to peer review and approval and publication of new books.

Technical competence pervades all SPE activities, not only in the publication of books but also in other areas such as sponsorship of technical conferences and educational programs. In addition, the Society publishes periodicals including *Plastics Engineering, Polymer Engineering and Science, Journal of Vinyl & Additive Technology,* and *Polymer Composites* as well as conference proceedings and other publications, all of which are subject to rigorous technical review procedures.

The resource of some 38,000 practicing plastics engineers, scientists, and technologists has made SPE the largest organization of its type worldwide. Further information is available from the Society at 14 Fairfield Drive, Brookfield, CT 10804, USA.

Michael R. Cappelletti
Executive Director
Society of Plastics Engineers

Technical Volumes Committee
Hoa Q. Pham, Chairperson
The B.F. Goodrich Company

Reviewed by:
Richard F. Grossman
Halstab Div. of Hammond Group Inc.

Preface

My original education was in chemistry, with a slant toward a career in research. After an undergraduate degree, a master's degree, and some subsequent work in graduate level programs, I had still never been exposed to Statistics and was in fact lacking in upper-level mathematical skills. Some people are fortunate enough to have a natural aptitude for math and acquire knowledge and skills in that subject almost intuitively. I am not one of those people and have spent more unhappy hours studying math in frustration and despair than are worth remembering.

Luckily, it is possible to become a competent chemist without being a mathematical genius, and once I entered industry and began practicing applied chemistry in polymer-related processes, my other training and skills proved to be more than adequate in letting me function well in both the laboratory and the factory.

Over the years my career progressed into management and positions in technical leadership, and eventually I underwent considerable instruction in subjects such as Deming training and Statistical Process Control. Because these subjects both involve heavy emphasis on statistical concepts and use of mathematical procedures, my education in mathematics was rekindled. I then accepted a post within a large corporation, in which there was access to a qualified statistician who had the particular advantage of being able to communicate in plain English to the statistically ignorant.

Over time I began to experience the effectiveness of statistical techniques in helping to understand what various collections of data might really mean or not mean, and, as our technical challenges grew in complexity, the usefulness of statistical techniques became more and more apparent. I began to see the tremendous value of such techniques when applied properly to the analysis of a number of very varied problems.

By that time I had heard of Design of Experiments, although we were not using them. The company brought in a local professor to teach a course on the subject which I attended with some difficulty and did not fully grasp, but it increased my interest in the subject. By then my friendly local statistician had left the company and moved away, so it was necessary for me to undertake more learning without his help. There followed some years of taking courses and seminars, reinforcing them with actual practice, and continuously studying a growing collection of textbooks.

Eventually, I did acquire reasonable competence in Statistics and then in the understanding of basic Design of Experiments methodology. But the learning had proved to be difficult, not only because of my less-than-outstanding math skills but also because of the difficulties resulting from varied and sometimes confusing nomenclature, contrasting approaches taken by different experts, and the tendency for many Statistics authors to write in a manner more suited to other statisticians than to run-of-the-mill scientists, engineers, and industrial workers. It was only by applying intense effort over a long time, with a great deal of learning by doing, that I was able to finally cut through the wording and contrasting philosophies to reconcile the underlying reality of various Design of Experiments methods and learn to use them effectively.

The good news is that basic understanding of designed experiments is not really that difficult to learn and use. The more advanced concepts and mathematics are certainly challenging, but it is possible to tackle the great majority of commonly encountered problems with only a few types of designs; and those designs are usually of workable size, easy to understand, and generate data that can frequently be analyzed and interpreted using nothing more than a handheld calculator.

Design of Experiments is a tremendously valuable tool for exploring new processes or gaining detailed comprehension of existing ones, and then optimizing those processes. It is a methodology widely employed in other countries, with a history of successes of such length and depth that it seems amazing that there has been such limited publicity about it and teaching of it in this country.

This comparatively narrow discussion is put forth in hopes of promoting the spread of learning about Design of Experiments. It is presented in what is intended to be a simplified and nonconfusing format and at a level where it can be immediately useful in itself, yet it can also provide a jumping-off place for those who may wish to move on to the many weightier publications available on the subject and particular schools of thought from individual authorities such as Drs. Box, Hunter, Taguchi, Wheeler, etc. The goals are to explain the basics underlying designed experiments, supply instructions on how to use several families of convenient designs that will be useful in very many situations, and provide some overviews on assorted subtopics of the large field that makes up Design of Experiments.

It is not necessary to read every chapter to gain from the book, because some are stand-alone summaries of specific subtopics. However, the early chapters (1–

10) do form the backbone of the presentation, and careful examination of them in turn is strongly recommended. Chapter 14 is also of special value.

Suggestions for improvements to future editions will be welcome, especially any that are related to actual situations and can be demonstrated using real data. May you find the book to be readable and useful, may all your data be precise, reproducible, and easy to interpret, and may good luck protect you from all the flaws and inaccuracies that experimenters so often encounter.

R.J. Del Vecchio
Southboro, Massachusetts

Contents

1   What Are Designed Experiments?

Every time someone sits down to consciously plan out some limited course of action to learn about what affects a process and how, that is in a real sense a designed experiment. When Dr. Walter Reed carefully exposed some volunteers to the bites of mosquitoes but not to any items or people associated with yellow fever, while giving other volunteers massive exposure to the soiled bedding and other items from those who had died of the illness but protecting them very carefully from mosquitoes, he had clearly planned out the experiment. (He could just as easily be said to have "designed" the experiment.) And his plan worked out well, establishing the mosquito as the true carrier of yellow fever.

The more effective scientists and technologists have always been able to plan and carry out experiments carefully so as to efficiently generate good data that reveal how things really work. Most of the time, however, they had to work through courses of experimentation in a series of small steps in each of which one factor at a time was changed in type or level of use (sometimes referred to as 1-FAT experiments). Proper experimental practice for 1-FAT work involves a lot of care to make sure that only the one chosen factor is changed during actual execution of the plan. This strategy can certainly work, and in fact the human race progressed from cave dwellers to living in skyscrapers by learning through simple experiments.

However, simple experiments can take a very long time to fully explore complicated processes, and even then attaining the best results depends heavily on the skill, past experience, and even the intuition of the individual experimenter. The more complicated the process being investigated or the more subtle the effects on the process are, the more likely it is for simple experiments to not yield all the desired knowledge.

Toward the end of the 18th century, mathematicians were exploring the analysis of certain patterns of numbers known as matrices. A Frenchman named Hadamard demonstrated that it was possible to extract a large fraction of the information in a matrix from a smaller fraction of the numbers in that matrix.

All the test results of all the possible experiments that could be run on a complex process, varying all the controlling factors at all their levels, would add up to a large group of numbers. This collection of numbers, each one of which is the result of a single possible experiment, can be considered to be a mathematical matrix. This means that running a fraction of all those possible experiments will still allow an investigator to learn almost as much as if he or she had run them all. This is not exactly getting something for nothing, but it is a way to get the biggest bang for the experimental buck.

One example would be a process controlled by four main factors, which could be feed, speed, cutting fluid, and alloy for a machining process; temperature, pressure, time, and concentration for a chemical process; or percentages of flour, sugar, milk, and eggs in a recipe. If each of the four factors could conceivably be used at just three different levels, all the possible experiments would add up to a total of 81, which is a lot for anybody to take the time to run. Yet doing as few as 16 of those 81 experiments, about one-fifth of the total, can reveal a great deal about the process; doing 25 experiments, less than one-third the total, can furnish almost as much information as doing all 81 might.

The key questions are, exactly which fraction of all the possible experiments will provide the desired information? and how can the data be analyzed so that the experimenter sees clearly what they reveal?

The most productive fraction of the potential experiments is determined by a mathematical breakdown of the full pattern, which then indicates just which subset of experiments needs to be run. A particular subset makes up an individual design. Over many years, mathematicians have worked out numerous separate designs or families of designs to fit different situations. Analysis of the data can often be done easily by uncomplicated methods but may also require more sophisticated, statistically based techniques.

Choosing the right fraction of many possible experiments to get the most information for the least effort and using whatever method is appropriate to properly analyze the results that come out of those experiments is what is meant by formal Design of Experiments. The field might be more accurately described as Statistically Optimized Experimentation, but it's now far too late to introduce a new term for this subject. (The name is very frequently abbreviated to initials, DOE or DOX; this writer prefers DOX, but DOE is somewhat more popular.)

2   Why and Where Should Designed Experiments Be Used?

The main reason for using designed experiments has already been explained (see Chapter 1). They are much more efficient than one-factor-at-a-time (1-FAT) investigations whenever more than one factor is thought or known to control a process. In addition, designed experiments will detect and quantify special relationships in which two or more factors act differently in how they affect process together compared to how they affect it separately. Such relationships are called *interactions*.

For instance, raising temperature alone might increase the yield of a polymerization reaction, whereas higher pressure alone might decrease yield slightly, but if hotter temperature and higher pressure together greatly increase the yield, that would be an interaction, in this case a positive one. Another example is two medications, each of which is good for a patient, but when both are taken together, a new, strong side effect such as severe nausea overcomes the patient. The two medications would be said to have a harmful interaction. Interactions do not always occur, but they are sometimes extremely important, and 1-FAT experiments by their nature are not capable of finding interactions.

Sometimes a factor affects a process in a nonlinear way, that is, the size of the effect of changing the factor level is not always proportional to the change in the factor. For instance, increasing the oven temperature from 300 °F to 325 °F might cut bread-baking time from 75 minutes to 50 minutes, but increasing the temperature to 350 °F only cuts the time to 40 minutes, 375 °F reduces it to 35 minutes, and 400 °F always burns the outside before the middle bakes properly. The line describing the response of the process to changes in the temperature would be a curve rather than a straight line. Use of the right design can quickly demonstrate nonlinear factor effects, typically in a way faster or easier than a series of 1-FAT experiments.

When a process has a long history and considerable expertise has been gained by those running it, or it is affected by only a few controlling factors and in comparatively simple ways, it is quite possible that a few simple experiments drawn up by people who know the process will serve perfectly well for fine tuning it; but when a process is new and has numerous likely control factors, certain classes of designed experiments can be remarkably efficient in rapidly determining which factors are most important, or which have a nonlinear influence on the process. These are basic screening designs.

If interactions are suspected for some factors in a process, then other types of designs can be used specifically to check for those interactions, or a design can be used that will detect any possible interactions. Finding and controlling interactions is often the key to getting the best results from a process, and when a process is a complicated one, subject to several control factors, their interactions, and some nonlinear effects, then the pattern of its responses to changes in factors can become quite complex. This is when a good designed experiment is absolutely invaluable, because it can provide really important information about the process at a cost and in a time greatly less than any other method. At such times designed experiments will give results that are unlikely or even impossible to obtain by even the longest series of simplified experiments.

Still another advantage of designed experiments is that it is possible to use the data to not only estimate what affects the process and how much but also to separate out effects which are really significant from those that are just meaningless numbers. This property of estimating the amount of normal scatter (or "experimental error" in the jargon of statisticians) in the test results is a special advantage of Design of Experiments (DOX) which can be of great value.

One other thing can be said about these kind of experiments: when a designed experiment has been properly set up and run, the experimenter *always* learns something of value. It may be how to optimize the process (the generally desirable point of the work) or that the process is simply not capable of doing what is wanted (not desirable, but better to know than not know), the choice of control factors was incomplete (again, better to know than not know), or perhaps the measurement system used for the testing was inappropriate. In all these cases, the work done is not wasted and important aspects of the process have been revealed. This is much better than the common experience of doing a lot of lab work and then winding up with results that are hard to interpret or cannot be used, or might even be misleading.

Although designed experiments are not automatically the only way to examine something, they very frequently represent the most cost- and time-effective method in situations in which any level of complexity or unfamiliarity with the process exists. This will become clearer in the coming chapters.

3 How Hard Are Designed Experiments to Use?

From one point of view, designed experiments are not hard to do at all. Consider a very possible case in which a new piece of injection molding equipment has been set up and needs to be started up. A brief team meeting reveals that it has eight different major control dials or buttons, which influence things such as stages of back pressure, injection pressure, holding pressure, runner temperature, molding temperature, cycle time, etc. There are also separate sources of material to be molded and possible pretreatments for the materials, so that the total number of variables in this process adds up to 10.

However, experimenting with ten things at once is very impractical. The classic way of starting up this machine would be to assign the most senior operator available to pick some initial combination of factor settings by best guess. Depending on how the process runs the first time, the operator would then make at least one change in a setting and run the process again, beginning a lengthy trial-and-error sequence.

With luck and a very skilled operator, perhaps it would only take 20–30 trials to hit on some combination that runs fairly well, but if Murphy's Law comes into play (a virtual certainty in any lab or factory), various interactions and nonlinear effects will prolong the process over many more trials, taking weeks to work through bad starts and dead-end approaches. Alternatively, the factor settings that do give fairly good results will turn out to be right on the edge of a stable process so that production winds up trying to run a delicate and erratic process that remains a source of problems.

Using Design of Experiments (DOX), one possible first step would be to use a screening design that contains only 12 different sets of conditions, each of which is referred to as an "experimental run" or "treatment." The screening design would separate the most critical two to five factors from all the rest and rate them in order of importance. Analyzing the test results would take less than 10 minutes, or only seconds if a simple computer spreadsheet function were used.

Let's say it turned out that four factors showed themselves to be crucial but the process was suspected to be subject to interactions, as injection molding often is. The simplest next step then would be to go to an interaction design, in this case a 16-run factorial that would detect and measure all possible interactions. (The other six factors found to be noncritical during the screening experiment would be kept at whatever levels were most desirable or convenient.) Again, analysis of the data would not be difficult.

The information from the two sets of experiments would have shown a reasonably clear picture of how the process works, and from that alone it is exceedingly likely that a set of conditions could be picked that would be at least as productive as what an expert operator might develop over weeks of trial and error. Which is easier, a few weeks of trial-and-error work covering 50+ different runs or a couple of days and 28 runs? And remember, the designed experiments will also provide the bonus of useful information on the basic normal scatter of the process. The use of some secondary techniques can also reveal whether the process is more variable or sensitive at some factor settings than at others.

If full optimization of the process were needed, a third experiment of a more advanced type could be used to zero in on the absolute best way to run the process. It might require 25 runs, but some of those might be identical to ones performed in the earlier work so that the prospect of being able to make fewer runs and still obtain the most accurate picture possible of the process would be good.

For most people, the data analysis for this type of experiment would tend to require a computer program, but some of those are not difficult to use. However, it often happens that the combined information from the two more basic designs is so effective in discovering how to control the process that this information alone is enough to get things running quite well. Subsequent fine tuning by operators or manufacturing engineers during normal operations can sometimes be very effective in completing the shakedown.

It has to be admitted that for many organizations and some individuals it still feels easier to go out and play with the process, doing one experiment at a time. Typically, the most difficult part of adopting DOX is the very normal resistance to change found everywhere. People will find a variety of things to object to, starting with, "You think we can take time to run 12 experiments? That's a lot of work to do when we're trying so hard to get production out, and besides, all that testing would put the lab into overload. Harry's had a lot of experience on this kind of process, why don't we just let him go out there and do his thing? He'll work it out soon enough."

Of course, this is the same type of thinking that says it is better to run at high volume even with excessive scrap levels so that shipments can be met, the same attitude that makes people feel they are too busy running the machine to take time to maintain or fix it. In the long run, such thinking and attitudes are very counterproductive and become far too costly to modern competitive businesses.

The "hard part" of using DOX is the necessity of carefully planning out what needs to be done, making the resources available to do it right, and then dotting every *i* and crossing every *t* to have the work done just as planned. This can take some real effort, but after all, performing technical tasks properly has always been a key part of science and technology, without which enormous amounts of work can be utterly wasted.

Learning a few types of designs and the uncomplicated graphing and math techniques for digesting the data is not terribly difficult. Such work does not require a degree in science or any degree at all. Technicians and workers with high school diplomas can and do plan and run designed experiments successfully every day across the world. Knowledge of working processes and a degree of common sense are more useful in practical experimentation than advanced knowledge of Statistics. Computer programs can be very helpful and convenient, but their use is not at all necessary to good use of basic DOX.

Designed experiments, like any other body of techniques, require some study and practice to become fully usable, but they are not really difficult to understand or use, and most capable people can acquire the needed knowledge and skills with only a moderate amount of effort. There is no reason for nervousness when beginning these studies; a generally positive attitude and the drive to acquire a very useful new skill are all a student needs.

4 Basic Statistics as Background to Design of Experiments

A *statistic* is any number measuring something, such as a person's age or height, a car's weight or horsepower, or a company's yearly sales or profit. The field of *Statistics* is a body of concepts and techniques for examining information, based on mathematically derived methods, which allows much better understanding of what the information really says. (This distinction is worth clarifying; statistics with the lower case s are just numbers, whereas Statistics is a subtopic in the field of Mathematics.) The two simple uses of Statistics are to help understand what has happened in the past and allow prediction or control of what will happen in the future.

By using statistical methods, it is possible to answer various kinds of questions, such as how wealth is distributed in some society, who is more likely to win the next election, whether a new experimental drug is really more effective than an existing medication, or which social trends represent minor fads and which are major changes in the way people think or interact. Often people are presented with a set of data to look over and try to understand, and the data set is really a collection of statistics, such as the individual heights of a group of people or all the household-income levels in a particular town. This really amounts to lists, sometimes long lists, of numbers describing some characteristic, such as height in inches or yearly income in dollars.

When there is a large collection of numbers, such as all the batting averages of players in a league, looking at them in one big group is not always immediately helpful to understanding what they mean. It becomes easier to make a graph of them. (Graphs and charts are always convenient in allowing people to examine a great deal of data in one glance and quickly grasp some idea of what it all means.)

The most typical way to graph a collection of data is to lay out a series of columns, with each column assigned to some number or small set of numbers from the larger full collection of numbers being examined. For instance, for people's weights, each column could contain all the weights within a 5-lb. range, such as everyone from 150 lb. up to 155 lb., and then for every person whose weight fell in that range, one X would go in the column. In this way the columns fill up with X's and the height of the filled column immediately shows how the number of people in that weight group compare to the number of people in any other weight group. This kind of vertical bar graph is called a *histogram*.

Figure 4.1 displays information on the height of a group of 169 men. Each column is assigned a 1-in. range of heights; for instance, the shortest column on the left in Fig. 4.1 is labeled 60 in. and indicates that only one man with a height of ~5 ft was found in this group. The biggest column is at 70 in. and shows that 24 men at 5 ft 10 in. were in the group.

The graph in Fig. 4.1 shows at a glance how the heights of the men are distributed among the group. The term "distribution" is used very frequently in statistics to describe the collection of numbers that hold information of interest. Note the shape of this distribution. It is something like a gentle hill or a bell shape, and it turns out that the majority of naturally occurring things have distributions of this general form. For this reason, this bell curve is usually called the "normal" distribution, although statisticians call it a

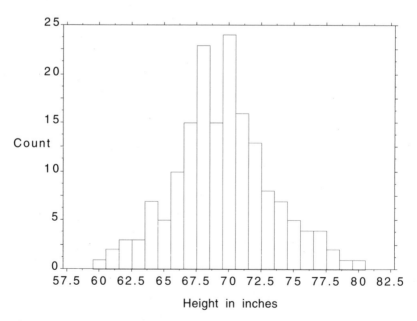

Figure 4.1 Histogram of heights of 169 men

Gaussian distribution because a mathematician named Gauss derived the equation that makes a smooth curve of that exact form.

The equation is a little complex, but the main thing about it is that only two terms or numbers can then be used to quickly characterize how whatever is being measured is distributed. These two numbers are the *average* of the distribution (which statisticians call the *mean*) and the standard deviation, which is a measure of how wide the distribution is.

It is easy to calculate the average of a group of numbers, and everybody knows what an average is. The standard deviation would take a bit more explanation and time to calculate, but fortunately many handheld calculators will now do it automatically. What is key to remember is that a normal distribution is approximately 6 standard deviations wide; that is, the distribution is centered on the average and the spread around the average is ±3 standard deviations (abbreviations such as Std. Dev. or even just SD are frequently used).

An ideal normal distribution curve is shown in Fig. 4.2, with its spread marked by standard deviations. The area under the curve represents all the members of the population, and it is easy to see that not much of the population falls in the intervals out at the two tail ends of the distribution. Because of the way normal distributions work, it is possible to define the amount of the population that falls in any particular interval of the distribution. Most of the time, however, people refer to three main intervals: : from −1 SD to +1 SD, containing 68.3% of the population; from −2 SD to +2 SD, containing 95.5% of the population; and from −3 SD to +3 SD, containing 99.7% of the population.

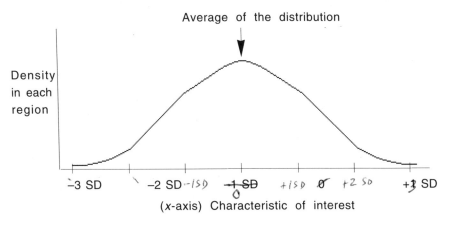

Figure 4.2 Ideal normal distribution

Ideally, distributions contain every single member of the overall group. This is called *the full population*, and in the case of men's heights, it would have to be for all the men in the state or country, rather than the 169 men who were measured. The 169 men are what is called a *sample group*. Most of the time we are examining samples instead of full populations.

Dealing with samples instead of full populations has some disadvantages, and a great deal of the application of Statistics involves the use of sample information to predict what the full population is. One example is taking a poll of 1,500 people to predict the manner in which many millions of people will vote in some election. The 1,500 individuals in the poll are a sample; everybody voting in the election is the population.

Often, getting sample data is difficult or expensive, and the question becomes one of how small a sample group can be used while still allowing judgments to be made about the population with confidence. One set of statistics techniques involves the calculation of the size a minimum sample group should be to allow satisfactory resolution of some question.

However, a rule of thumb is that really high confidence begins with sample groups of 100 or more, and reasonable confidence begins with groups as small as 30. Obviously much smaller samples (2–6) are often used or have to be used in some comparisons and can still lead to good answers if properly analyzed. Other statistical techniques deal with how to do proper analysis of limited data, but it is always preferable to maximize the amount of test information as best you can. Designed experiments work in part because they are a way to increase the amount of information about the effects of each included control factor (see the explanation of "hidden replication" in Chapter 6).

Only a few distributions commonly encountered are truly nonnormal, but many are only roughly normal. They might be skewed to one side or the other in shape, have two peaks instead of one, or be cut off sharply on one or both sides (see examples in Fig. 4.3).

When a distribution is not fully normal, there are some problems in using standard statistical procedures, which is why distributions are often drawn into histograms and checked for being seriously out of shape. If that happens, then statisticians will use various special techniques to analyze the distribution and still be able to use the information. Fortunately, such concerns do not often apply in designed experiments, so it is not necessary to learn too much more about how to deal with nonnormal distributions.

One major use of the SD is to determine the percentage of any group that falls into some defined subgroup. Recall that, in a normal distribution, 68.3% of the total group falls in the ±1 SD interval around the average, 95.5% falls

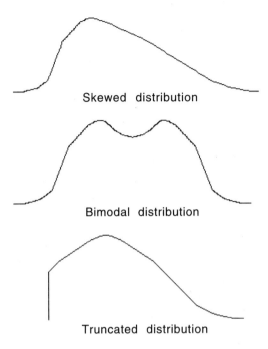

Skewed distribution

Bimodal distribution

Truncated distribution

Figure 4.3 Examples of distributions demonstrating nonnormality

in the ±2 SD interval, and 99.7% falls within ±3 SD. Thus if you had a distribution of individual product parts that averaged 100 g in weight and had an SD of 2 g, you would know that one-half of the parts weighed between 100 and 106 g. (That is the average plus 3 SD, which has to be one-half of the distribution.) If someone asked how many parts weighed 104 g or more, you would know that about 2.25% of them fell in that class (~4.5% of the distribution is outside of ±2 SD, which is from 96–104 g, so only one-half of that is >104 g, which is 2.25%.)

In statistical analyses and comparisons, the idea of probability has to be understood, because the kinds of conclusions that can be reached are almost always probability statements. For instance, if two groups of test data from different batches of material are compared and appear to be neither identical nor very different, the question would be, are these two batches really significantly different from each other?

However, the answer would not be a clear yes or no. Instead, it might be something like "If you assume the two batches are different, there is a 10% chance you will be wrong." This naturally implies that there is a 90% chance they are different, which is called significance at the 90% confidence level.

What level of confidence is enough to satisfy someone about making a judgment on something? That depends on the situation and the person. Sometimes an 80% confidence level is sufficient to allow an action to be taken; other times 95 or 99% confidence is required. (If the decision is to risk $10 on buying something to make your car run better, the 80% may sound good; if the decision is to risk your own life on the use of an experimental operation, 99% sounds a lot better.)

The validity of such probability statements depends partially on the techniques used (and statisticians debate extensively on the worth of different techniques) but very heavily on the quality of the data available for analysis. When poor or questionable data are used, the probability statements that then result still read the same way but can be anywhere from less accurate to seriously misleading.

Design of Experiments (DOX) methodology is one way of generating very high-quality data in minimum quantity so that various means of analysis, including such tests as analysis of variance and regression analysis (discussed in later chapters), can be done with high levels of confidence in the results.

One of the more significant principles in Statistics is the Central Limit Theorem (CLT). It is a useful mathematical principle that can be formally derived, but that level of academic demonstration is not essential to its understanding or application. The two major points of the CLT are as follows:

1. The averages from a set of samples of any population will be distributed approximately normally around the population average (whether or not the parent population is normally distributed).
2. The SD of the population of sample averages (called the *standard error*) will be smaller than the SD of the parent population by a factor of $1/\sqrt{n}$ where n is the size of the sample groups.

The larger n is, the more normal the distribution of the sample means will be. This is useful in providing reassurance that experimental data derived from averages can be manipulated with standard techniques, because those techniques depend in part for their validity on being used with data from normal distributions. If the type of original test data is suspected or even known to be highly skewed or otherwise deviated from normality, as long as it is averages from sample groups that are being examined and compared, the many convenient standard procedures for analyzing and comparing data can be used with confidence.

Also, the larger n is, the more narrowly distributed the sample averages will be around the true population average. This means that if a range of values (called an *interval*) is calculated so as to contain the true average, that interval becomes more and more narrow. For instance, suppose the average of a set of samples tested for tensile strength is 4,000 psi and the SD of the sample is 150 psi, and there were 3 pieces in the sample. The standard error of the test (which is the same as the SD of the distribution of the sample averages) would be 150 psi/1.732 ($150/\sqrt{3}$), or 86.6 psi.

Because ±2 SDs cover 95% of the spread of the distribution of the sample averages, and that distribution is centered on the mean of the parent population, the real tensile strength of the material is ~95% certain to fall in the range of 4,000 ± [2(86.6)] psi, or 4,000 ± 173 psi, which calculates out to a range from 3,827 to 4,173 psi.

The 3,827–4,173 psi range is called a *confidence interval*, and in this case it is the 95% confidence interval for the tensile strength of the material, based on a sample size of 3.

However, if the sample size had been 9 instead of 3, then the standard error would have been 50 psi and the 95% confidence interval would have shrunk to 3,900–4,100 psi. The use of confidence intervals is common in several kinds of statistically based comparisons, including those made in designed experiments.

The important points to remember about basic Statistics are:

- many things fall into normal distributions, and if they do, various good techniques for analyzing the data can be applied;
- the average and SD are the numbers most often used to conveniently characterize a distribution;
- the more data (larger sample) you have, the better, and the more you know about how the information was gathered, the better;
- the use of averages from sample groups makes analysis more secure, because the CLT shows that even if the analyzed property is not distributed normally, the averages will be;
- larger samples will more accurately indicate where the true average of the population lies, with confidence increasing with the square root of the sample size.

Even limited exposure to the overall subject of statistics would require a much more substantial presentation than is appropriate or even possible in this context. This narrow look is intended only to help build some understanding of the underlying principles of designed experiments and the analysis of the resulting data. For a deeper understanding of those principles, especially as applied to more advanced DOX practice, the inquiring student is referred to any of the excellent texts listed at the end of this book.

5   Fundamentals of Experimentation

Design of experiments (DOX) is basically the use of particular patterns of experiments to generate a lot of information about some process while still using an absolute minimum of actual experiments to get the information. Before going into what those patterns are, how they work, and how best to use them, it is necessary to understand two things: first, some background on interpreting the results of any set of experiments; and second, the basics of correctly drawing up and performing any set of experiments.

It is principal fact that variation exists everywhere and that if we use a measurement system that is sensitive to small differences, we will not get identical readings for a series of things measured, no matter how careful we are to make the things as alike as possible. Whether identical twins or a bunch of peas in a pod, when things are measured carefully enough, differences will be detected.

If a process is run many times as identically as possible and the results are properly measured, a collection of readings will then be found. These readings can then be considered a distribution and be examined like any distribution. They will almost always fall into a normal distribution, with its own average (mean) and width [measured by the standard deviation (SD)].

When experiments are run to explore what it takes to make some process output change, the experimenter is looking to see what factor or cause does produce a noticeable difference in how the process runs and what it produces. However, by the natural law of normal variation, no process makes exactly the same product every single time, so the experimenter has to expect at least some differences in output every time the process runs, whether or not the factor being explored really affects the process.

The question then becomes, how much of a difference does there have to be between the output of experimental runs before it is safe to conclude that the difference demonstrates a real change in the process and is not just part of normal variation? Or, put another way, if we test a set of supposedly identical specimens and get a range of results, what number in that range is the "real" result for that type of specimen in that test?

The answer involves another concept from Statistics, which was introduced in Chapter 4 and is called the standard error. (It has nothing to do with making mistakes but is actually an estimate of the normal amount of variation involved in the output of the process.)

All samples that are made or tested are meant to relate to some real or theoretical population, and it is the average and SD of that population which are being sought. If only a single data point is taken, it can fall anywhere in the parent distribution and by itself contains no information on the SD of the population, but if a sample group is taken, its average falls somewhere in another distribution, called the *sampling distribution*.

The sampling distribution is the distribution of all the possible sample averages for the particular sample size chosen. The sampling distribution is much larger than the parent distribution, because the number of combinations of samples that can be taken from a distribution is always considerably greater than the number of individuals within it.

What is important is that the CLT shows two things. First, because the sampling distribution is made up of averages, it has to approach normality even if the parent distribution is not normal, and its average will very closely approach the true average of the parent distribution. Secondly, its SD will be related directly to the true SD of the parent population (called *sigma*, symbolized by σ) by the inverse square root of the sample size, N. This is stated mathematically as

$$\sigma_x = \sigma / \sqrt{N}. \tag{5.1}$$

The term σ_x is the SD of the sampling distribution, but it is also known as the standard error. Because the sampling distribution is normal, 95% of the points in it will fall within ± 2 SD of its center point, and that center point is both the *grand average* of all sample averages and the best estimator possible of the true population average.

When a sample group is generated, its average ($\overline{X}$) is the only estimator available for the grand average of all samples (which is the estimator for the true population average). If the standard error were a known quantity, the statement could be made that the sample average had a 95% chance of being within ± 2 standard errors of the grand average, but calculating the true standard error by the equation above requires the use of sigma, which is itself unknown.

Because certain knowledge of sigma is lacking, the only alternative is to use whatever estimator of sigma is available, in this case the standard deviation (s) of the sample group, so the estimated standard error is

$$\text{standard error} = s / \sqrt{N}. \tag{5.2}$$

This applies to the interpretation of real data as shown in the example in Chapter 4; when a sample group is tested for some property, the interval ±2 standard errors around the property mean is considered to have about a 95% chance of containing the true level of the property of interest. (There are detailed statistical techniques to more precisely adjust the width of the confidence interval, but they will not be explored here.)

In actual experimentation, the measured results from two different sets of experimental runs must not have their confidence intervals overlap or they cannot be judged to be really different from each other. If the intervals only overlap partially, it is quite possible for the two results to actually be from different populations, but there are just not enough data to support that conclusion at a high level of confidence. Answers are often not "yes" or "no" in Statistics but "probably yes" or "probably no" instead.

The amount of normal variation in test results for test specimens made in what is intended and set up to be the same process is commonly called the *scatter* of the process. Every process is subject to normal variation and every process has its own level of scatter, but what contributes to scatter, and why is it broader for some process than for another?

We can subdivide scatter into two broad categories, *bias scatter* and *random scatter*. Bias scatter occurs when there is a pattern of the data being inaccurate. For instance, if a thermocouple reads wrongly, having been set 10° too high, all the temperatures will be off by 10°. If in the course of some operation the equipment steadily gets hotter than its set point over time, the temperatures will be all off, but in clearly increasing order. Perhaps in the middle of an experiment, someone changes the lots of raw material so that one-half of the experiment is run with the old batch and one-half with the new batch. In all these cases, the difference in the experiments is in some sort of regular pattern, as illustrated in Fig. 5.1 (next page).

Note that the temperatures noted by x, all on the high side, are not perfectly even anyhow. The kind of unpatterned up-and-down variation seen there and in the points represented by • and * as well, is called *random variation*.

We know that it is impossible to eliminate variation and patterns of variation altogether, but we want to absolutely minimize it. That makes it much easier and more valid to associate changes in the data with real changes in the process. Four main defenses can be used against scatter when running experiments.

The first defense is quite basic and is simply to be as careful and methodical as possible in controlling all the circumstances of the experiments. This is referred to as *scientific rigor*, and although it is easy to understand the idea, it is often not so easy to put it into practice. It means being meticulous about

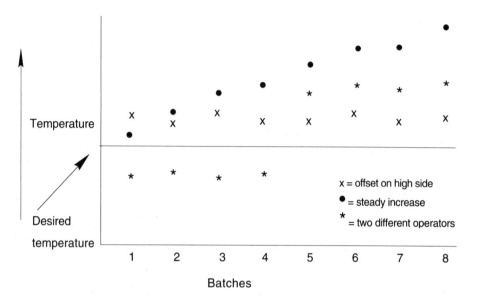

Figure 5.1 Types of bias scatter

every detail, weighing ingredients precisely, controlling or monitoring temperatures carefully, avoiding all sorts of contamination, running things as much as possible with the same people on the same equipment at the same time, etc., and generally having complete directions drawn up that are followed as written. It sometimes means checking and double checking on exactly what was done and how it was done. It means convincing everyone involved in drawing up and actually running the experiment that they need to devote their full attention and careful thought to every step of the experiment.

The second line of defense is the use of randomization when running a series of experiments. The pattern of experiments as drawn up is very structured, but they do not have to be run in the order in which they were drawn up. In fact, running them in any kind of ordered sequence should be avoided as much as possible. There are many ways to randomize a set of experiments, including the simple method of putting slips of paper representing each experimental setup in a container and then taking them out one at a time (like a lottery); the individual experiments are then run in the order their slips were picked out of the container. By randomizing the runs, the chances are greatly decreased for having some sequence effect, such as equipment heating up during the day, build in a bias that gives false indications of what affects the process. Also, because randomization is easy enough to do, on paper at least, it becomes a standard part of most experimental work.

Blocking, the third defense, is the method of subgrouping experimental conditions so that comparisons within the subgroups or blocks become more precise. For instance, suppose an experiment involved both changes in ingredients and two different mixing procedures. The various batches could always be mixed in pairs, each having the same ingredient combination but using the two mixing procedures consecutively. That would be a paired design with a block size of two. The data from the experiment would then be particularly useful in determining any difference in effect from the two mixing procedures.

Sometimes blocking is used not to bring more precision to an experiment but to help deal with a problem in implementing a design. For example, if it takes a long time for processing temperatures to change and stabilize, running experiments with the temperature randomized could pose substantial difficulties. If all the runs at one temperature were done first and then all at the second temperature, the comparison of temperature effect would be indistinguishable from a comparison of the order in which the runs were made in time.

This might be dealt with by making one-half of the lower temperature runs first , then one-half of the higher temperature runs, then the remaining low temperature runs, and, finally, the last of the high temperature runs. Within each block, the runs would still be randomized. In this way, the temperature is only changed three times to give four blocks of data. Temperature effect is still estimated by contrasting results of low versus high, but now any effect of time order of the runs has been diluted and can be examined separately by comparing average effects of each of the four blocks.

The fourth, last, and most powerful defense against scatter is to generate data more than once, preferably several times, at one or more of the experimental conditions. Running the same experimental setup more than once is called *replication*, and replication is used whenever practical in good experimental work. Examining the scatter from results of replicated setups provides a good measure of scatter for all the results of the several similar sets of conditions that make up a designed experiment.

The replications are normally done at intervals between the other experiment runs (even if they have been randomized). However, if the entire set of experiments is replicated, as is sometimes done, they may be done in two complete sets, which is another form of blocking.

Once the scatter in the data is understood and measured (whether it has been minimized or not), it is easier to make judgments about how much meaning can be found in the data. This leads us to the concept of the *signal-to-noise ratio*, which is important in understanding what the data do tell us (and also in knowing what kind of experimental design we might need to use).

The "signal" referred to is the change you are looking to see; for instance, if you are trying to listen to someone's voice, the pattern of sound they make is the signal you want to receive. In an experiment, the change will be in the level of some response, such as the tensile strength of some material.

The "noise" is the background variation related to the kind of signal. In the example of someone's voice, the noise is literally the background noise in the room. In the case of tensile strength, the noise is the normal scatter in tensile strength of the material.

If the noise is small compared to the signal, it is easy to get good reception. A loud speaking voice in a quiet room is easy to hear and understand. However, if the noise is large compared to the signal, for instance, a regular speaking voice (a moderate signal) in the middle of a rock concert (a very large noise), it becomes very difficult to receive the signal clearly.

In another example, if a plastic's tensile strength data had a scatter of ±500 psi but the kinds of changes being made in the material were only likely to raise tensile strength by 100 psi, it would be hard to find changes of 100 psi.

This is why it is really important to understand both the background scatter (noise) of whatever process is being investigated as well as how noticeable the effect (the signal) is that you are hoping to find. A favorable ratio is one that is fairly high, for instance, a signal that is 4–10 times larger than the noise. If this is not the case for the property being measured, then certain techniques for picking out experimental designs can come into play.

In preparing for any series of experiments, a certain set of procedures is often very effective. In simple outline it has only a half-dozen steps:

1. Have a clear understanding of what the problem is or what question needs to be answered, and, on that basis, develop a basic statement of the goal of the experiments. This principle may seem obvious, but it is not always respected in practice the way it should be.

2. Have a meeting of everyone who is concerned with the situation, including those who know the problem, those who are working on the solution, and those who will be running the experiments and doing the testing. Start off with a general discussion of the situation and the goal of the experiments. Include discussion of what responses should be examined to check on changes in the process.

3. Use the brainstorming technique to list all the possible factors that are at all likely to have any effect on the process. Once that is done, list the factors in rough order of suspected importance, and get a consensus on which should be part of the experiment.

(Some may be considered too minor to bother with, and some may be important but can be set at one level and left out of these particular experiments.)

4. Move on to consider what range of the control factors should be part of the experiment. The general idea is to be a bit bold and cover a range of factor levels that is pretty sure to make a difference in the process, but that applies differently to different kinds of factors. For some process that normally runs at 200 °C, a range of 180–220° is bold (which looks like ±10%), but the range of some ingredient might go from 1% to 3%, which is ±50%, seemingly much larger. This is where the past experience of some people, plus what others have heard or read, plus some common sense, can add up to a reasonable best guess.

5. Consider the number of factors and their levels, pick the appropriate experimental design, and then plan out in full detail exactly how to run it. Again, use the knowledge and experience of everyone in checking out what the problems might be in performing the work, in getting the testing done, and in getting everything done most efficiently with the least chance of things going wrong. The plan must be documented, written down clearly (perhaps in the meeting minutes, perhaps in a special format), with assigned actions for everyone who will be involved.

6. Lastly, the experiment has to be done as planned. All the best meetings and discussion and planning in the world finally boil down to a collection of people with specific tasks to do and a schedule for doing them. This final step is where things sometimes break down.

Unfortunately, to get all the desired benefit of DOX, which is a great deal of first-class information from a minimum of experiments, it is very important to run the pattern as planned. Missing even a small part of the pattern or running it differently from the plan can cause a major loss in the quantity and/or the quality of the data. Thus a certain amount of focus or devotion to plan, or even stubbornly hanging on to make it happen, is important to have in the people who are involved in the work.

One extra step that is strongly recommended is to have a review session later on with the whole group about the work and the results. The more everyone knows about what was learned and how well the plan worked, the more experimentation skills will be developed in the organization.

Of course, it is quite possible for a single individual to act as a "team of one" in considering a situation, examining variables and their appropriate levels, planning out the experimental design, and perhaps even carrying out the experiments and doing the testing. Many famous scientists worked that way. However, a team approach in experimentation tends to be more effective overall.

The major points of this chapter are:

- Variation occurs normally in every process, so even if the process has not been changed by varying some factor, there will be some detectable difference between experimental runs;
- knowing how much normal process variation (scatter) there is becomes important to make it possible to judge when a real change has occurred in the process;
- the comparison between normal scatter (noise) and the change you hope to see (signal) is also important, and the larger the signal is compared to the noise, the better the chance of learning about the process;
- the two kinds of scatter are the results of bias and random chance;
- the best ways to minimize scatter include being very careful (scientific rigor), randomizing the experimental runs, using blocking when appropriate, and most especially using repeated conditions (replication);
- the best way to run designed experiments includes knowing what you are looking for, getting inputs from all the people who can help, making a good plan, and then doing the work just as planned.

6 Basics of Experimental Designs

In examining any process, what is being sought is a detailed understanding of the relationship between things that can be changed in the process and the effects on the output of the process that result from changing those things. Any condition or setting that can vary during a process is called an *independent variable* by statisticians, and any output or result of the process is termed a *dependent variable*. Independent variables are commonly referred to as control factors and become symbolized by X, and dependent variables are known as responses, characteristics, results, etc., symbolized by Y.

Some control factors can be varied continuously, such as temperature, time, etc., whereas others only fall into singular classes, such as one ingredient compared with another or the absence of a step versus its presence in the process. The second kind of factor is described as being discrete, as opposed to the continuous type. Whether factors are continuous or discrete can affect what type of experimental design is used in investigating a process. Responses are usually continuous.

Control factors are changed systematically during experimentation, and whether continuous or discrete, their different settings are referred to commonly as *levels of use*. In a set of experiments, each separate set of conditions is called a *run* or *treatment*.

The most basic pattern of experiments would be to use every single possible combination of control factors and levels. This is called a "full factorial" experiment. For instance, suppose someone wants to experiment with two process temperatures and two process times, say 350 °F versus 400 °F and 4 min versus 6 min. The calculation for how many experiments are in the full factorial is made by multiplying the total number of levels of the first factor by the number of levels of all other factors. In this example, we have two factors at two levels, so the equation becomes:

(Levels of A) × (Levels of B) = Total experimental runs

$$\text{or} \quad 2 \times 2 = 4. \quad\quad\quad (6.1)$$

The actual pattern (which is formally called a matrix) is shown in Fig. 6.1

350 °F 4 minutes	350 °F 6 minutes
400 °F 4 minutes	400 °F 6 minutes

Figure 6.1 Basic 2×2 designed experiment

As the number of factors and their levels grows, the total of experimental runs goes up very quickly. If there were four factors at three levels each, the equation becomes:

$$3 \times 3 \times 3 \times 3 = 81. \tag{6.2}$$

In drawing up experiments, it is a common practice to designate the factor levels by some code instead of using them written out all the way. If there are two levels, as in the example above, they are usually coded as -1 and $+1$ levels, and in the matrix sometimes only the sign ($-$ or $+$) is used. Because two-level experiments make up the main part of simple DOX methods, we will focus on that technique in this chapter.

For example, if we had three factors (A, B, and C) and wanted to run every combination of $+$ and $-$ levels possible, the list would look like that in Fig. 6.2.

A little work will show that no other combination of $+$ and $-$ levels exists for three factors. (The order in which the combinations are shown is not significant; the run numbers can be reversed or randomized without changing anything of importance.)

	A	B	C
Run 1	+	+	+
2	+	+	−
3	+	−	+
4	+	−	−
5	−	+	+
6	−	+	−
7	−	−	+
8	−	−	−

Figure 6.2 Fundamental pattern of a 2-level, 3-factor full-factorial design

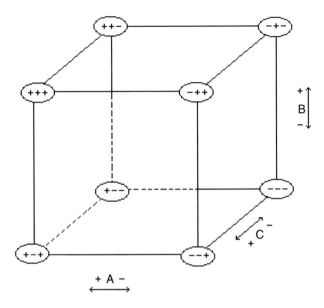

Figure 6.3 Three-dimensional representation of 3 × 2 full-factorial design

The pattern corresponding to this list is a cube in space, with each of the corners representing one of the factor combinations. This is easier to see in the actual diagram of (Fig. 6.3).

Note in Fig. 6.3 that each corner has a combination of + and – signs in it. Each sign corresponds to one of the runs in the list in Fig. 6.2 The volume contained in the cube represents the "experimental space," which is the many combination of conditions that fall between the eight particular combinations we selected for the experiment.

The arrows in Fig. 6.3 show how each factor is held at one value, + or – , for one face of the cube. Thus the cube surface closest to the viewer is the + setting for factor C, the right hand plane is the— setting for A, and the top plane is the + setting for B.

Other types of patterns can be drawn, such as a cube with experiments not only at all the corners but also in the middle of each plane and one more in the center of the cube itself. Those are the more advanced designs mentioned in Chapter 3.

An important concept in DOX is called *hidden replication*. Remember that making a judgment on some property is always much, much safer when the property is evaluated by averaging several experimental results than when only one measurement is used. This is applied in DOX by comparing the

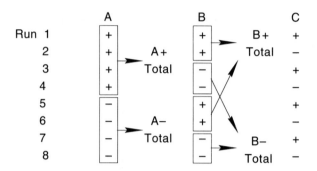

Figure 6.4 Hidden replication in 3-factor, 8-run pattern

average of some response property measured from all the runs made at one set of conditions to the average of all the runs at the other condition.

Looking at the 3-factor, 8-run pattern again, we can see in Fig. 6.4 that, for factor A, the first 4 runs are at the + level and the second 4 runs are at the – level. The average property from runs 1–4 can be compared to the average property from runs 5–8, and this gives us more confidence that when we see a difference between the averages, it is really telling us the process if affected by going from A+ to A– conditions.

In the same way, we can look at Fig. 6.4 for factor B, where we compare the average of runs 1, 2, 5, and 6 to the average of runs 3, 4, 7, and 8. This principle is applied to any factor in any formal design.

In the 8-run experiment, we get a bonus, so to speak, of checking out three factors instead of just one, with four replicates at each of two levels. This is what is meant by hidden replication, and it is a key facet of designed experiments.

Is that all the information that can be extracted from the eight runs? Statistical theory says that from eight independent pieces of data (in this case, the eight measurements of the response characteristic that result from performing the experiment), seven independent comparisons can be made. Because the experiment will tell us about factors A, B, and C, four other comparisons can be made.

These comparisons can be made by again averaging four + responses and comparing this average with that of four – responses, but by a particular pattern only. The additional four columns of + and – assignments are shown in Fig. 6.5, labeled as X1 through X4.

Note that if the pluses and minuses are added up in any column, the total cancels out to zero, and if the settings in any one column are multiplied by the corresponding settings in any other column, the resulting pluses and minuses

	A	B	C	X1	X2	X3	X4
Run 1	+	+	+	+	+	+	+
2	+	+	−	+	−	−	−
3	+	−	+	−	+	−	−
4	+	−	−	−	−	+	+
5	−	+	+	−	−	+	−
6	−	+	−	−	+	−	+
7	−	−	+	+	−	−	+
8	−	−	−	+	+	+	−

Figure 6.5 Fully expanded 3×2 pattern

again total up to zero. This relationship is called *orthogonal*, and experimental designs that fit this description are sometimes termed *orthogonal arrays*. (Not every conceivable experimental design is orthogonal, but the most efficient designs are typically orthogonal.)

These four columns can be used primarily in two different ways. One way is to assign four more control factors to the experiment, so that seven potential influences on the process are being evaluated. The X columns then become factors D, E, F, and G. At this point, the design has become a "fractional factorial," and because all columns have been assigned to independent factors, it is called saturated. (An 8-run design containing only three factors is a full factorial, but if it has between four and six factors, it is called an unsaturated fractional factorial design.)

Saturated designs are often used with good success in evaluating a set of comparatively unknown factors, not all of which are believed to be important to the process. The technique permits determination of which of them are really significant and which are not. The design is often called a screening design, because it separates important from unimportant factors. Such patterns are very efficient for that purpose; a full factorial for seven factors at two levels would require $2 \times 2 \times 2 \times 2 \times 2 \times 2 \times 2 = 128$ runs, which is 16 times as many as in the saturated screening design.

However, it is still true that we cannot get something for nothing, and by using saturated designs we automatically give up all possibilities of learning about anything other than the main effects of the control variables. Yet it is certainly possible for factors to affect each other's effects on the process rather than acting totally independent of each other; that is, temperature and pressure could both speed up some chemical reaction rate separately, but when both are increased, the speed of the reaction might go up geometrically.

This phenomenon, when it exists, is called a *factor interaction* and may be of strong interest in a given process. Saturated designs do not permit any judgments about possible factor interactions, but this may be not be a problem depending on the process being investigated or the state of the investigation.

The second way of using the unassigned columns is to not add any other factors to the experiment. In that case, it is possible to look for interactions between factors. This is done by looking at the product of the signs of factor settings in the first three columns. For instance, if A in each run is multiplied by B in that run, the first run yields a + sign and the second run yet another, but the third through the sixth runs yield – signs, and the last two yield + signs again. (Look at the array in Fig. 6.4 and run through this yourself.) That exact pattern of + and – is seen in column X1.

This means that when the average of those runs marked + in column X1 is compared with that of those marked –, the effect being examined is that of any possible interaction between factors A and B. X2 is then the A–C interaction, X3 is the B–C interaction, and X4 represents the three-way A–B–C interaction. If no interaction exists, the + and – averages from the appropriate column will be close to each other. (Theoretically, the + and – averages would be equal to each other when there is no interaction, but the normal random variation in any process almost always leads to some minor difference between the averages.)

In many processes, factor interactions are not common, and the use of saturated designs is very appropriate and highly efficient. However, for chemical processes and many mechanical/chemical systems, interactions are regularly encountered, and the use of a saturated design can result in false or misleading data. Estimation of whether interactions exist is one of those areas where the combined experience and judgment of several people familiar with the process may turn out to be helpful. When in doubt, the best course is to assume interactions are likely to be there and draw up the experiments accordingly.

As was already stated, theory tells us the differences between averages within columns representing an unimportant factor or nonexistent interaction should be zero, but we know that by random chance each column's average will vary a little in its value, so very few of the differences will really show up as zeros. So how do we decide which column differences represent real effects and which are just results of normal random noise?

There are special mathematical techniques for making that judgment, but fortunately a simple technique called a *scree plot* is quite effective and very easy to use. It is often sufficient for determining what factors/interactions are important enough to carry on to the next stage of the investigation.

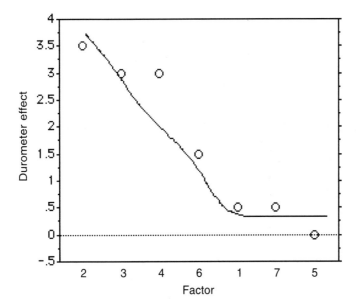

Figure 6.6 Scree plot of durometer effect

A scree plot is made by examining the effect levels seen in the data and ranking them by their absolute size. Their sign is not important in this ranking, so an effect of −3 is larger than one of +2. The scale of the graph has to be just a bit larger than the largest effect. The effects are plotted by size across the graph. Normally, a curve of some sort results, as in the example in Fig. 6.6 from an experiment on effects of ingredients on hardness of a rubber compound. Because there were seven factors evaluated in a saturated 8-run screening design, they are identified by number at the bottom of the chart.

This plot can be thought of as a side view of a cliff, going down to the base level of ground. The rubble of rock debris usually found at the base of a cliff is called *scree*, which is where we get the name of the plot. Where the leveling off of the "data cliff" can be seen is where the assumption can be made that significance of the effects shown starts to become very questionable. In Fig. 6.6, factors 2, 3, and 4 have substantial effect on rubber durometer, factor 6 has a lesser effect, and factors 1, 7, and 5 probably do not have any real influence on durometer.

Another technique is to examine the value found for a column assigned only to a three-factor interaction. It is rare for three-way interactions to be of any significance, so often the effect level observed in such a column can be considered to be representative of pure scatter. If that is true, then any other effect levels seen that are of comparable size are also probably results of normal scatter rather than any real effect of a factor or interaction.

Major points to remember are:

- Full factorial patterns involve every factor at every level and permit detailed analysis of effects and interactions;
- designs reduced in size from full factorials are called fractional factorials;
- using the averages of results from all runs set at the + or – levels of the factor being checked gives greatly improved confidence that the averaged effect is real;
- two-level designs are often tremendously useful and convenient;
- if interactions may be important, they can be analyzed using empty columns in unsaturated designs;
- if interactions are not important, saturated designs can be used with high efficiency;
- scree plots are useful in comparing relative effects of factors, etc.

7   Fractional Designs and Their Uses

Most of the time, experimenters prefer to avoid using large full-factorial designs, so the development of numerous types of fractional factorials has been a major part of the modern history of design of experiments (DOX). To understand when and how to best use the various designs, it is important to grasp the basics of what fractional designs really are, and it is easiest to begin with the most basic fractional, the half factorial.

This is exactly as the name implies: the use of only one-half the number of experimental runs called for by a full factorial. The halves have to be split up in a balanced manner. For instance, Fig. 7.1 shows the same three-factor full-factorial design used previously in Chapter 6 but with each run designated as belonging to the "right" or "left" half.

Figure 7.2 shows the two separate half designs (which are termed complementary to each other) side by side. The bar between them can be thought of as the plane of a mirror, each half design being a reflection of the other, with the mirror reversing the values or signs of the run conditions. Both designs contain the same information and will support the same type of data analysis.

Because there are four experiments in one of these half factorials, there is enough information to allow three judgments to be made from the data. (A

Factor	A	B	C	(Right or Left)
Run 1	+	+	+	R
2	+	+	−	L
3	+	−	+	L
4	+	−	−	R
5	−	+	+	L
6	−	+	−	R
7	−	−	+	R
8	−	−	−	L

Figure 7.1 Basic 3×2 design with right and left halves designated

A	B	C			A	B	C
+	+	+	R	L	−	−	−
+	−	−	R	L	−	+	+
−	+	−	R	L	+	−	+
−	−	+	R	L	+	+	−

Figure 7.2 Complementary half factorials

A	B	C	Interaction =	AB	AC	BC	ABC
+	+	+		+	+	+	+
+	−	−		−	−	+	+
−	+	−		−	+	−	+
−	−	+		+	−	−	+

Figure 7.3 3 × 2 half factorial with interaction patterns

statistician would say there are three *degrees of freedom* available; degrees of freedom are just a measure of information, and usually if there are *n* experiments or pieces of data, there are *n* − 1 degrees of freedom.) In this case, those judgments would be the effects of the three control variables A, B, and C, which would be seen in comparing the average response from the runs with A at its + level versus the average of the runs with A at its − level. This is the same technique used for the full eight-run design.

In the eight-run design, however, it was possible to also separate out the interactions. To do that, we multiplied the signs for the interacting columns and used the resulting pattern to check on the interaction effect. When that procedure is applied to one of the half factorials, we see the interaction in Fig. 7.3.

Notice that the AB interaction column is identical to the C factor column, the AC column is the same as the B factor, and the BC column is exactly like the A column. This means that any effect from the AB interaction is completely mixed in with the main effect of the C control factor. (Formal DOX terminology is that the effects are *confounded* or *aliased*.) If there is an AB interaction, its real effect is found using the exact same set of response numbers used to find the main effect of the C factor, and the two effects (from C and AB) cannot be separated from each other.

This is what happens in a saturated fractional factorial; every main factor column has confounded with it some interaction(s). With the half factorial, the confounding is comparatively simple, but it usually becomes much more complex with other fractional factorials.

For instance, an eight-run design with seven factors used will have three separate two-factor interactions confounded with every main factor effect. (There are also interaction terms involving three or more factors, but those are generally ignored in this kind of work.) In an unsaturated design, some factors may not have as much confounding associated with them, but it takes careful examination to know where interaction terms fall in the pattern.

Notice that the ABC interaction column in Fig. 7.3 only has + levels in it. This means that no judgment at all can be made about the interaction, but the fact that there are no – terms tells us that it was the ABC column of the original eight-run design that was used to split the eight runs into the two half factorials. By using the [ABC = +] runs for the right half and the [ABC = –] runs for the left half, a perfectly balanced split was made that resulted in each remaining column having one factor and one interaction in it. This is a form of blocking, with the ABC column being used as the blocking key.

The main point of all this is to show that whenever screening designs are used, interactions are confounded and that the effect measured for some main factor may in reality be due partially or largely to some irresolvable interaction of other factors. As mentioned earlier, this may not be important for some types of work, but when significant factor interactions are possible or even likely, screening designs must be used with care.

Among the types of designs available are a set called Plackett-Burman (P-B) designs, invented about 1939 and used with great effect in the World War II British war effort. Among the P-B designs are a subtype called nongeometric designs, which are the 12-, 20-, and 24-run designs. In these designs, the interaction contributions are spread out over the columns instead of falling totally in one column or another. Thus the nongeometric patterns can be used to screen main effects in systems where the existence of interactions is suspected, with much less chance of getting misleading data on main effects. (Interactions would then be explored in second-phase designed experiments.) Two of the designs are shown in Figs. 7.4 and 7.5 (next page).

12-run	A	B	C	D	E	F	G	H	I	J	K
Run 1	+	+	−	+	+	+	−	−	−	+	−
2	+	−	+	+	+	−	−	−	+	−	+
3	−	+	+	+	−	−	−	+	−	+	+
4	+	+	+	−	−	−	+	−	+	+	−
5	+	+	−	−	−	+	−	+	+	−	+
6	+	−	−	−	+	−	+	+	−	+	+
7	−	−	−	+	−	+	+	−	+	+	+
8	−	−	+	−	+	+	−	+	+	+	−
9	−	+	−	+	+	−	+	+	+	−	−
10	+	−	+	+	−	+	+	+	−	−	−
11	−	+	+	−	+	+	+	−	−	−	+
12	−	−	−	−	−	−	−	−	−	−	−

Figure 7.4 Plackett-Burman 12-run design

20-run	A	B	C	D	E	F	G	H	I	J	K	L	M	N	O	P	Q	R	S
Run 1	+	+	−	−	+	+	+	+	−	+	−	+	−	−	−	−	+	+	−
2	+	−	−	+	+	+	+	−	+	−	+	−	−	−	−	+	+	−	+
3	−	−	+	+	+	+	−	+	−	+	−	−	−	−	+	+	−	+	+
4	−	+	+	+	+	−	+	−	+	−	−	−	−	+	+	−	+	+	−
5	+	+	+	+	−	+	−	+	−	−	−	−	+	+	−	+	+	−	−
6	+	+	+	−	+	−	+	−	−	−	−	+	+	−	+	+	−	−	+
7	+	+	−	+	−	+	−	−	−	−	+	+	−	+	+	−	−	+	+
8	+	−	+	−	+	−	−	−	−	+	+	−	+	+	−	−	+	+	+
9	−	+	−	+	−	−	−	−	+	+	−	+	+	−	−	+	+	+	+
10	+	−	+	−	−	−	−	+	+	−	+	+	−	−	+	+	+	+	−
11	−	+	−	−	−	−	+	+	−	+	+	−	−	+	+	+	+	−	+
12	+	−	−	−	−	+	+	−	+	+	−	−	+	+	+	+	−	+	−
13	−	−	−	−	+	+	−	+	+	−	−	+	+	+	+	−	+	−	+
14	−	−	−	+	+	−	+	+	−	−	+	+	+	+	−	+	−	+	−
15	−	−	+	+	−	+	+	−	−	+	+	+	+	−	+	−	+	−	−
16	−	+	+	−	+	+	−	−	+	+	+	+	−	+	−	+	−	−	−
17	+	+	−	+	+	−	−	+	+	+	+	−	+	−	+	−	−	−	−
18	+	−	+	+	−	−	+	+	+	+	−	+	−	+	−	−	−	−	+
19	−	+	+	−	−	+	+	+	+	−	+	−	+	−	−	−	−	+	+
20	−	−	−	−	−	−	−	−	−	−	−	−	−	−	−	−	−	−	−

Figure 7.5 Plackett-Burman 20-run design

Taguchi designs are also used frequently, but they are mainly variations on P-B designs and other classic designs, using different notation and some extra calculations not liked by formal statisticians. However, they work well, and some of the ideas Dr. Taguchi has made popular about experimentation and process optimization have proven to be very useful.

With any screening design, an alternate set of runs can be made using the reflected design; that is, with all settings reversed. Just as with the reflected half factorials seen above, the two designs will contain the same information; selection of one design over another can be made on any convenient basis, including whichever set of runs is easier to perform.

However, it is occasionally very useful to run both sets of experiments from a screening design, which means all the original and reflected combinations of factor settings. This doubles the number of experimental runs needed but is still far smaller than a full factorial for all the factors. When this is done, it becomes possible to get some information on two-factor interactions from the dual array, which is called "folded over."

So far, the discussion has centered on experiments in which the factors are only used at two levels, because such experiments can easily be learned and used, yet are still extremely useful in many situations. However, when a continuously variable factor is used at just two levels, only linear measurements of its effect can be made. (Discrete factors, such as different suppliers or machines, etc., are always considered nonlinear.)

In many kinds of systems, factor effects are nonlinear, especially if the range of factor levels is broad. To measure nonlinearity, at least three levels of a factor are needed. This leads to more complex patterns of experiments and more complicated ways of analyzing the data; this subject will be introduced in Chapter 15.

Nonetheless, two-level designs are used for the bulk of the designed experiments run day to day in factories and laboratories. They are easy to draw up, run, and analyze, and they often reveal very important information about processes. In many cases, simply knowing which factors are the most significant in a process and whether or not interactions are important is enormously helpful to understanding a process and how to more efficiently gain control of it.

Major points of this section are:

*   Use of fractional-factorial designs always leads to confounding of interaction effects with main factor effects;
*   location of interaction effects in an unsaturated fractional factorial may be useful or even necessary at times;

- if screening of several control factors is needed and interactions are suspected but will be neglected for the moment, use of nongeometric Plackett-Burman designs is desirable;
- reflected designs contain equivalent knowledge and can be used whenever convenient;
- fold-over designs can be useful in locating interactions;
- two-level designs, although limited in capacity to profile nonlinear effects, are generally both useful and convenient.

8   Examples Using Eight-Run Design

Even though the concepts of basic two-level designs and the arithmetic needed to analyze the resulting data are not difficult at all, actual use of the patterns still requires some practice beforehand. Three examples of the eight-run design will be used to illustrate in detail how they can work.

In the first case, only three factors are under consideration, which means the eight-run design represents a full factorial. The pattern of the levels of the three factors (A, B, and C) is shown in Fig. 8.1, along with the resulting data from the individual experimental runs. In this case, the response is shown as Y and could represent the percentage yield of some reaction.

The factors could be time, temperature, and pressure, or levels of three catalysts or monomers, or any three factors affecting the reaction. An actual worksheet might look like the array in Fig. 8.2.

Because this is an example of a full factorial, the interaction column patterns can be used, which leads to the matrix shown in Fig. 8.3.

In the next step, the Y responses corresponding to the – level in the A column are totaled up, and then the + responses are totaled separately. Each total is entered in a separate row below the A column, and then the total of the – data points is subtracted from the total of the + data points to give the difference between the two subgroups; this difference is then divided by four

	A	B	C	Y
1	–	–	–	65.3
2	–	–	+	81.3
3	–	+	–	53.3
4	–	+	+	69.9
5	+	–	–	61.8
6	+	–	+	77.4
7	+	+	–	73.9
8	+	+	+	89.9

Figure 8.1 An 8-run design with coded levels of factor

	Temperature, °F	Time, h	Catalyst Conc., %	Yield, %
1	200	1	1	65.3
2	200	1	2	81.3
3	200	2	1	53.3
4	200	2	2	69.9
5	250	1	1	61.8
6	250	1	2	77.4
7	250	2	1	73.9
8	250	2	2	89.9

Figure 8.2 Pattern of design in Fig. 8.1 as applied to a typical worksheet with one response

	A	B	C	AB	AC	BC	ABC	Y
1	−	−	−	+	+	+	−	65.3
2	−	−	+	+	−	−	+	81.3
3	−	+	−	−	+	−	+	53.3
4	−	+	+	−	−	+	−	69.9
5	+	−	−	−	−	+	+	61.8
6	+	−	+	−	+	−	−	77.4
7	+	+	−	+	−	−	−	73.9
8	+	+	+	+	+	+	+	89.9

Figure 8.3 Pattern of Fig. 8.1 expanded to show interaction columns

	A	B	C	AB	AC	BC	ABC
+ Total	303.0	287.0	318.5	310.4	285.9	286.9	286.3
− Total	269.8	285.8	254.3	262.4	286.9	285.9	286.5
Difference	33.2	1.2	64.2	48.0	−1.0	1.0	−0.2
Effect on Y	8.30	0.30	16.05	12.00	−0.25	0.25	−0.05

Figure 8.4 Calculation of effects from response data in Fig. 8.3

(the number of experiments that made up each subgroup) to provide an estimate of the average effect on process yield of changing the A factor from its − to its + level.

This procedure is repeated for each column in the matrix, which leads to the data display in Fig. 8.4. (Notice that the − total and + total always add up to 572.8, which is the total of all eight Y values.)

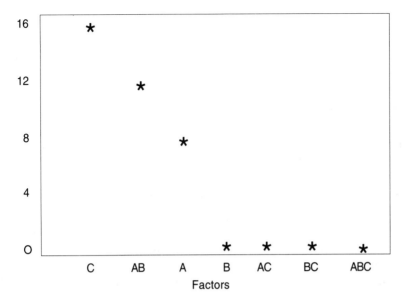

Figure 8.5 Scree plot of effects on process yield

In the example used, these data demonstrate that going from the lower temperature to the higher will, on average, increase the yield by about 8%, whereas increasing the catalyst concentration has twice as much of an effect. Time by itself has very little effect, but in conjunction with temperature does contribute to a substantial increase in yield, as shown by the effect shown in the AB column of Fig. 8.4.

The scree plot of these effects (Fig. 8.5) will show a dramatic contrast between the A, C, and AB effects, with all the other column effects falling close to each other and to zero.

This is a comparatively simple example of use of the fully unsaturated eight-run design in which the data showed clear contrasts. Use of saturated or partially saturated designs allows consideration of more factors in the experiment but can lead to increased complications.

The most effective way to use an eight-run design when interactions are suspected between some but not all of the factors is to set up a four-factor half factorial design, with the fourth factor settings located in the ABC interaction column. The basis of this plan is the common (but not infallible) assumption that two-factor interactions are much more often both real and of significant magnitude than are higher level factor interactions.

For instance, assume that in the example above the three factors were ones which could easily interact but there was also interest in using a larger reaction vessel. General belief among those knowledgeable in the process is that although reaction vessel size might affect yield caused by differences in agitation levels, any interactions between the vessel and the other three factors would be very unlikely. The set of experimental runs could then be as shown in Fig. 8.6, with factor D being reaction vessel size.

Analysis of the second data set supplies a new set of estimated effects, but no real changes are seen for the effects of temperature, time, catalyst concentration, and the temperature/time interaction, although there appears to be a minor effect of reactor size in lowering yield (Fig. 8.7).

The use of half factorials when at least one control factor is known or strongly believed to be independent of all others is a common and effective practice. It is possible to use more complicated, partially saturated fractional designs when a mixture of several noninteractive and interactive factors is involved, but this can become both difficult to lay out properly and sensitive to any misjudgment on the existence of interactions.

The use of fully or almost fully saturated designs for screening purposes only is a major part of DOX strategy. An example of six factors in an eight-run design is shown in Fig. 8.8.

This experiment involved effects of ingredient levels on properties of a rubber compound. It is a reflected design (as shown by the row of all minus

	A	B	C	D (= ABC)	Y
1	−	−	−	−	65.6
2	−	−	+	+	79.3
3	−	+	−	+	51.3
4	−	+	+	−	69.6
5	+	−	−	+	59.8
6	+	−	+	−	77.7
7	+	+	−	−	74.2
8	+	+	+	+	87.9

Figure 8.6 Half-factorial design with data related to design in Fig. 8.1

	A	B	C	AB	AC	BC	D
+ Total	299.6	283	314.5	307	282.5	282.9	278.3
− Total	265.8	282.4	250.9	258.4	282.9	282.5	287.1
Difference	33.80	0.60	63.60	48.60	-0.40	0.40	-8.80
Effect on Y	8.45	0.15	15.90	12.15	-0.10	0.10	-2.20

Figure 8.7 Calculation of effects from response data of Fig. 8.6

values) with the last column unused so as to act as an estimator of the amount of random error in the process. Two responses are shown, the first relating to the minimum viscosity (ML) of the unvulcanized rubber and the other relating to the delay in beginning of vulcanization reactions after processing temperature has been reached (T2). The analysis of the first response is shown just below the pattern of the experimental runs in Fig. 8.8.

The data indicate that, as expected, addition of a second, low viscosity polymer does decrease the material viscosity significantly, whereas the third, fourth, and sixth factors all increase viscosity. The remaining factors do not appear to affect this property significantly and would show up on a scree chart as the group closer to the zero-value line.

When the analysis is extended to the second property (Fig. 8.9), a different pattern of results is observed.

	LVP	MS1	MS2	MS3	PrsAid	Xlnkr	G	ML	T2
1	+	+	+	−	+	−	−	16	3.4
2	+	+	−	+	−	−	+	15	3.4
3	+	−	+	−	−	+	+	19	3.4
4	+	−	−	+	+	+	−	17	4.5
5	−	+	+	+	−	+	−	24	2.9
6	−	+	−	−	+	+	+	18	3.8
7	−	−	+	+	+	−	+	20	3.4
8	−	−	−	−	−	−	−	14	5.7
+ Total	67	73	79	76	71	78	72		
− Total	76	70	64	67	72	65	71		
Difference	−9	3	15	9	−1	13	1		
ML effect	−2.25	0.75	3.75	2.25	−0.25	3.25	0.25		

Figure 8.8 A six-factor eight-run fractional factorial design with calculation of effects of ML response. LVP, low viscosity polymer; MS1–MS3, metal salts or oxides; PrsAid, process aid; Xlnkr, chemical cross-linker; G, unused factor; ML, minimum viscosity; T2, delay in beginning vulcanization reactions until after processing temperature is reached

	LVP	MS1	MS2	MS3	PrsAid	Xlnkr	G
+ Total	14.7	13.5	13.1	14.2	15.1	14.6	14
− Total	15.8	17	17.4	16.3	15.4	15.9	16.5
Difference	−1.1	−3.5	−4.3	−2.1	−0.3	−1.3	−2.5
T2 effect	−0.275	−0.875	−1.075	−0.525	−0.075	−0.325	−0.625

Figure 8.9 Calculation of T2 effects (second response) from Fig. 8.8

The second and third factors (metal salts or oxides) act as catalysts in the cross-linking reaction between the polymer and the chemical cross-linker, so their contribution to the delay in reaction initiation was anticipated to be negative, i.e., they would shorten T2. This is exactly what the data indicate, along with lesser effects from the fourth and sixth factors. The fifth factor was a process aid known to be chemically unreactive, and its effect on T2 is very close to zero.

However, the unused column (also called vacant, dummy, or empty) that was supposed to estimate process scatter has an apparent effect level greater than that seen for columns four and six. This could be interpreted to mean that there is a great deal of scatter in the experiment and that most of the seeming effects are subject to question, even though they do fit in with prior knowledge of the process. However, there is another explanation.

In partially saturated geometric designs, the various possible interactions are confounded with main effects. Even though the last column is empty of any main factor, it can contain two-factor effects. In fact, in this case it contains the negative of both the MS1–MS2 and MS3–Xlnkr interactions. (Readers should refer back to Chapter 7 and go through the comparison to confirm this.)

Because both of those interactions contain elements that are chemically active and could certainly interact, the G column is probably not acting as a valid estimator of experimental scatter in this instance. The fifth column is a much more likely indicator of experimental error. How to more accurately examine the extent of scatter will be dealt with in a later chapter.

The main points of this chapter were:

- Simple computational methods can be used to analyze two-level designs, and
- the use of varied screening designs can be easy and powerful but also subject to some pitfalls.

9 Effective Use of Simple Designs

From the academic viewpoint, experimentation is not limited by its potential degree of difficulty, but in actual work with designed experiments, a certain degree of pragmatism usually becomes necessary. Although there are reports in the literature of even 256-run experiments having been performed, the vast majority of designs used contain fewer than 32 runs. The most frequently used designs are screening or simple interaction types ranging from 4 to 20 runs, and it is these easy-to-use and easy-to-understand designs that are the main focus of this book. We will examine them all in turn, comparing and contrasting how they can be used.

The four-run design can be used in two ways:

1. as a full factorial for two factors at two levels, and
2. as a half factorial for three factors at two levels.

(As always, the ability to recognize and evaluate interactions is lessened in the fractional factorial; in this particular case, it is utterly lost.)

The eight-run pattern, which is the most used of all designs, can be used in three ways:

1. As an interaction model for three factors at two levels,
2. as a half factorial for four factors at two levels (main effects stay unconfounded with two-factor interactions), or
3. as a fractional factorial for five to seven factors at two levels (but two-factor interactions become confounded with main effects).

The nine-run design can be either of two quite different types. It could be a full factorial for two factors at three levels, which provides information on both interaction and nonlinear effects, or it might be the Taguchi L9 array, which is basically a three-level screening design for up to four factors; in this design, information is generated on nonlinear effects, but no analysis of interactions is possible.

The two patterns are contrasted with each other in Fig. 9.1. (The classic convention of factor settings is -1, 0, and $+1$, and this is used in both patterns. Dr. Taguchi prefers coding two-level experiments as 1 and 2 and three level

Two factors, three levels: Full factorial		Four factors, three levels: Taguchi L9			
Factor	A	B	A	B	C	D
Level	−1	−1	−1	−1	−1	−1
	0	−1	−1	0	0	0
	1	−1	−1	1	1	1
	−1	0	0	−1	0	1
	0	0	0	0	1	−1
	1	0	0	1	−1	0
	−1	1	1	−1	1	0
	0	1	1	0	−1	1
	1	1	1	1	0	−1

Figure 9.1 A 2 × 3 full-factorial design compared with a Taguchi L9 design

experiments as 1, 2, and 3, but it would probably be confusing to mix both types of coding here.)

Notice that the first two columns of each pattern are identical, just switched in order. The extra two columns in the L9 are both orthogonal to all the others, which is normal for valid experimental design patterns.

The simplest way to analyze data from three-level screening designs is to graph the average result of each level versus the other levels for each factor in the matrix. If no obvious visual contrast can be seen between the three averages, then that factor is unlikely to be exerting any significant effect on the process. (The level of difference between the highest and lowest of those three averages may be used as an indicator of process scatter.) If there is a marked difference between the highest and lowest of the three averages, the pattern in which the averages fall can be interpreted so as to reveal what type of linear or nonlinear effect that factor has on the process and how large the effect is.

For instance, assume that the four factors in an L9 are catalyst type, reagent concentration, temperature, and pressure and that the reaction percentage yields observed for the runs are as shown in the worksheet in Fig. 9.2.

The analysis of the data then would involve taking the averages of each subgroup of three according to the levels of each factor. For instance, for Catalyst type, the first three runs used a nickel catalyst and averaged a 61.8% yield, the three silver catalysts averaged a 62.7% yield, and the platinum catalysts averaged a 69.5% yield.

If these calculations are done for all the control factors, the resulting numbers come out as shown in Fig. 9.3.

Catalyst	Concentration, %	Temperature, °C	Pressure, bar	Yield, %
Ni	20	60	0.5	57.9
Ni	25	75	1	64.6
Ni	30	90	5	63.0
Ag	20	75	5	62.3
Ag	25	90	0.5	61.7
Ag	30	60	1	64.1
Pt	20	90	1	65.5
Pt	25	60	5	68.1
Pt	30	75	0.5	75.0

Figure 9.2 Worksheet for a Taguchi L9 design

Level	−1	0	1
Catalyst	61.8	62.7	69.5
Concentration	61.9	64.8	67.4
Temperature	63.4	67.3	63.4
Pressure	64.9	64.7	64.5

Figure 9.3 Calculated average responses for three-level experiment

Most of the time, it is more convenient to examine the data in the sort of combined graph for all four factors illustrated in Fig. 9.4.

The response line for pressure appears to be declining but in fact is so shallow (range of 0.4% in yield) that it could easily fall within the spread of random variation instead of reflecting any real effect on the process. In contrast, the range of yield with reagent concentration is 5.5%, which is more than large enough to be significant. The pattern of the concentration effect appears to be largely linear.

The three different types of catalyst display differences, with the third type (platinum) standing out clearly as promoting higher yields than the other two. Because this is a discrete factor rather than a continuous one, there is no point to analysis of the shape of the response curve.

Finally, the response to temperature is dramatically nonlinear, showing a peak response at the intermediate level. In the case of a continuous variable such as temperature, it becomes possible to speculate on the mechanism underlying the response pattern, such as competing reactions at higher temperature which convert materials into alternate forms and thereby decrease yield.

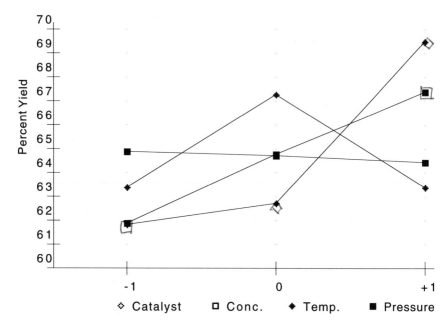

Figure 9.4 Effect chart for Taguchi L9 design

More sophisticated forms of analysis can be applied to L9/L18 data, but much of the time they will not provide any better insight into the process than the comparatively easy and understandable graphing method.

The 12-run nongeometric Plackett-Burman pattern is very useful for screening up to 11 main factors (at two levels) without any two-factor interactions confounding main factor effects.

The 16-run design can be used

- As an interaction model for four factors at two levels,
- as a fractional factorial for up to eight factors at two levels with no confounding of two-factor interactions and main effects,and
- as a fractional factorial for up to 15 factors at two levels (with confounding of main effects and interactions).

The classic 16-run pattern is shown in Fig. 9.5, except that columns 11–14 are identified with four possible extra factors instead of the three-factor interactions from factors A–D and the last column is identified with one possible two-factor interaction instead of the ABCD four-factor interaction.

If the four extra factors are actually used in the design, the two-factor interactions fall in the columns in Fig. 9.5 as follows:

Run	A	B	C	D	AB	AC	AD	BC	BD	CD	E	F	G	H	AH
1	+	+	+	+	+	+	+	+	+	+	+	+	+	+	+
2	+	+	+	−	+	+	−	+	−	−	+	−	−	−	−
3	+	+	−	+	+	−	+	−	+	−	−	+	−	−	−
4	+	+	−	−	+	−	−	−	−	+	−	−	+	+	+
5	+	−	+	+	−	+	+	−	−	+	−	−	+	−	−
6	+	−	+	−	−	+	−	−	+	−	−	+	−	+	+
7	+	−	−	+	−	−	+	+	−	−	+	−	−	+	+
8	+	−	−	−	−	−	+	+	+	+	+	+	−	−	−
9	−	+	+	+	−	−	−	+	+	+	−	−	−	+	−
10	−	+	+	−	−	−	+	+	−	−	−	+	+	−	+
11	−	+	−	+	−	+	−	−	+	−	+	−	+	−	+
12	−	+	−	−	−	+	+	−	−	+	+	+	−	+	−
13	−	−	+	+	+	−	−	−	−	+	+	+	−	−	+
14	−	−	+	−	+	−	+	−	+	−	+	−	+	+	−
15	−	−	−	+	+	+	−	+	−	−	−	+	+	+	−
16	−	−	−	−	+	+	+	+	+	+	−	−	−	−	+

Figure 9.5 A 16-run design shown with 8 factors

Column 5 = AB = CE = DF = GH
Column 6 = AC = BE = DG = FH
Column 7 = AD = BF = CG = EH
Column 8 = AE = BC = DH = FG
Column 9 = AF = BD = CH = EG
Column 10 = AG = BH = CD = EF
Column 15 = AH = BG = CF = DE

As pointed out above, eight factors could be screened in this design with no confounding of any of them with two-factor interactions. However, having four two-factor interactions in each remaining column would make any analysis of such interactions impractical.

The 18-run Taguchi design is useful when more than four process factors can be or need to be evaluated at three levels but some other factor (possibly discrete) is only to be varied over two levels. The 18-run design uses one factor at two levels and up to seven factors at three levels. As with the L9, information on interactions cannot be extracted from the data, but nonlinearity is readily apparent.

The 20-run, nongeometric Plackett-Burman pattern is used just as the 12-run, but it can screen up to 19 main factors instead of 11. Because it is very seldom that >15–19 factors need to be evaluated for a process, this is about the largest screening design that sees regular use.

Thus a group of seven different sizes of experiments (4, 8, 9, 12, 16, 18, and 20 runs), with seven more subtypes related to the way they might be used, is readily available to the investigator. How the choice made to employ one over the others depends on what is known of the process being examined.

When very little is known, interactions are possible or suspected, and a moderate number of factors (5–19) are involved, the goal of the experiment might well be to simply sort out the major factors from the minor ones. In such cases, the 12- and 20-run, nongeometric Plackett-Burman designs are the first possibilities that would come to mind.

In the less common situation in which significant effects from interactions are believed to be very unlikely, the eight-run array could be used to screen up to seven factors.

In screening experiments, the question often arises as to whether information on interactions or the nonlinearity of responses is more important. When two to four factors are involved and nonlinearity is of major interest, the L9 pattern is preferred, but if interactions may be more important, the four-run design is good for two factors and the eight-run design will handle three or four factors. (Of course, for only two factors there is the option of using a full factorial at three levels to generate complete data in nine runs.)

The L18 array can be used to screen up to seven factors in a three-level design, but, just as with the L9, no information on interactions will be available. The extra facet of one more factor at two levels can be used for an actual process factor or some alternative such as blocking two days of experimentation, two test machines, two lots of raw material, etc.

A 16-run, two-level factorial could be used to check the same eight factors for general significance, but then all the two-factor interactions would be confounded so thoroughly that it would be extremely difficult to attribute any effect in an interaction column to a particular interaction. Thus the eight factors might be more efficiently examined by the Plackett-Burman 12-run design.

However, if at least two of the factors were known to not take part in interactions, it would become possible to set up the column assignments so that more valid location of interactions became reasonably probable. If four of the factors were noninteractive, complete resolution of the remaining interactions would then be easy.

The question of interaction probability is a primary consideration in choosing and fine-tuning experimental designs. This why past experience with a given process and its likely interactions is so important to any DOX strategy and why inputs from a variety of individuals who have process familiarity can be key to developing a successful strategy.

In summary, the investigator taking the first-stage look at a process has to determine the following to select an appropriate design:

- How many factors are thought to be possibly significant to the process? and how small a subset of those factors needs to be examined initially?
- Should the goal of the first design be simple screening for factor significance, or should interactions/nonlinearity be checked?
- If interactions/nonlinearity are significant topics, which is the more important at this stage?
- If interactions are of key interest, what factors might safely be assumed to not be involved in interactions?

Once the answers to these questions are available, the choice of an initial design (which might be part of a series of designs) might proceed fairly easily, as long as the size of the most appropriate pattern still fits within the constraints of time, materials, etc., available to the experimenter. If that is not the case, rethinking of the immediate goal or reordering of priorities may be the answer, or adaptation of designs to fit the situation.

For instance, suppose the original goal had been to investigate five main factors, complete with interactions, but limitations on time or money make an experimental series larger than an eight-run prohibitive. If one of the factors can be judged less important or more easily fixed at one level temporarily, a four-factor half factorial will fit in the remaining factors and allow at least some chance of finding interactions.

Alternately, if time is not available in the week for an uninterrupted set of experiments but the need for good data is crucial, having the work done on a weekend may be the answer and an appeal to higher management for authorization might be appropriate.

If there are four factors to be evaluated at three levels and two others at two or three levels, but only six experiments can be done in a single day, a difficult problem arises: $3 \times 3 \times 3 \times 3 \times 2 \times 2 = 324$ runs in the full factorial, so clearly a fractional factorial is needed. However, six runs per day may add in too much scatter from other time-related effects, so the design needs to have some way of evening out the day-to-day variation.

In this case, the L18 would offer a very workable solution. The factors that might be run at two levels are reset for three levels, which means all six factors would be uniformly at three levels. The L18 array could be used with the remaining unassigned three-level column as the blocking pattern for three days of experiments. The unused two-level column becomes available as an estimate of scatter.

Key concepts of this chapter include:

- A variety of uncomplicated small to medium designs (4–20 runs) exist that can be used for different purposes;
- choice of a design depends on what knowledge is being sought, such as main factor effects only or interactions as well, or nonlinear effects; and
- knowledge of the various designs coupled with adaptability on the part of the experimenter can overcome many challenging situations.

10   Application of Folded-Over Designs

It has been stressed repeatedly that, in geometric screening designs, there is a danger of observing an effect which, although it appears to be a result of the particular factor assigned to that column, is actually the result of an interaction of two other factors. This often causes an experimenter to resort to use of the nongeometric Plackett-Burman designs, which will unambiguously identify main factor effects free of interactions.

With the nongeometric designs, however, separating out interaction effects is not practical even if that becomes desirable, and sometimes the size of the nongeometric design is larger than people want to use. For instance, using a 12-run experiment for only five factors or a 20-run for nine factors seems inefficient to many investigators.

There is an alternate strategy made possible by the known consequence of having interaction effects change sign when a reflected design is used. This means that seven factors could be evaluated in an eight-run design, which would then show some contrast in effects for at least a few of the factors. If a serious question were then raised as to which of the observed effects might actually be caused by interactions, it would only be necessary to run the reflected eight-run design and examine the new data in comparison to the old.

Every factor that showed an effect of about the same magnitude as seen in the first set of runs, and with the same sign, could then be judged as being the true cause of that effect. Those factors whose effects remained roughly the same in size but changed in sign would be revealed as not really being related directly to the effect seen. Those effects that reversed sign between the two sets of runs would then be shown as the results of interactions.

This can be illustrated with experiments in rubber compounding in which tensile strength was one response of interest. The factors to be evaluated were two different types of polymer, carbon black reinforcement, mineral filler, sulfur level (cross-linker), curative (cross-linker and reaction initiator), accelerator (strong reaction initiator), and process aid. The first saturated eight-run is shown in Fig. 10.1.

Factor	Polymer	Black	Clay	Sulfur	Cure	Accel	PrsAid	Tensile Strength
Run 1	–	–	–	+	+	+	–	2,105
2	–	–	+	+	–	–	+	2,014
3	–	+	–	–	+	–	+	2,350
4	–	+	+	–	–	+	–	2,325
5	+	–	–	–	–	+	+	1,971
6	+	–	+	–	+	–	–	2,128
7	+	+	–	+	–	–	–	2,309
8	+	+	+	+	+	+	+	2,439
+ Total	8,846	9,424	8,906	8,867	9,022	8,840	8,774	17,641
– Total	8,795	8,217	8,735	8,774	8,620	8,802	8,867	
Difference	51	1,207	171	93	402	38	–93	
Effect	12.8	301.7	42.7	23.3	100.6	9.5	–23.3	

Figure 10.1 A saturated eight-run fractional factorial with one response; Cure, curative; Accel, accelerator, PrsAid, process aid

In this case, it appears that the major factor in tensile strength is the carbon black, which would be expected from copious past experience. Clay and sulfur would be expected to also increase tensile strength, and the process aid might be expected to decrease it slightly, so the data appear reasonable in those cases. However, the second greatest increase in strength falls under the curative column, which would not be expected to have a larger contribution than sulfur, the main cross-linker.

A reflected design was set up and carried out, with results as shown in Fig. 10.2.

Note that the effect levels are reasonably comparable for all the columns except column 5, which has gone from a magnitude of +100 to –62. This change in sign indicates that what is being observed is not the direct effect of the curative but some interaction.

The combined data sets from both the original and reflected designs is called a folded-over design, and the analysis of all the data together by the normal method provides valid indications of the real factor effects independent of the interactions. When this is done with the two data sets above, the analysis line is represented in Fig. 10.3.

In Fig. 10.3, the true effect of the curative can be seen to be about equal to that of the sulfur, which fits far better with the history of this system. This still adds up to a total of 16 runs for seven factors, but even a quarter-fac-

Reflection	Polymer	Black	Clay	Sulfur	Cure	Accel	PrsAid	Tensile Strength
Run 9	+	+	+	−	−	−	+	2,386
10	+	+	−	−	+	+	−	2,315
11	+	−	+	+	−	+	−	2,137
12	+	−	−	+	+	−	+	2,000
13	−	+	+	+	+	−	−	2,352
14	−	+	−	+	−	+	+	2,367
15	−	−	+	−	+	+	+	2,030
16	−	−	−	−	−	−	−	2,056
− Total	8,805	8,223	8,737	8,787	8,946	8,795	8,860	17,643
+ Total	8,838	9,420	8,906	8,856	8,697	8,848	8,783	
Difference	33	1,197	168	70	−249	53	−77	
Effect	8.4	299.3	42.0	17.4	−62.2	13.3	−19.3	

Figure 10.2 Reflected design of Fig. 10.1 with data

	Polymer	Black	Clay	Sulfur	Cure	Accel	PrsAid	Tensile Strength
+ Total	17,685	18,844	17,812	17,724	17,719	17,688	17,557	35,284
− Total	17,600	16,440	17,473	17,561	17,565	17,596	17,727	
Difference	85	2,404	339	163	153	91	−170	
Effect	10.6	300.5	42.4	20.3	19.2	11.4	−21.3	

Figure 10.3 Calculation of combined responses from both original and reflected designs

torial for seven factors requires 32 runs, so a substantial increase in efficiency has been gained in finding specific main effects while at least determining the existence and extent of a significant interaction.

Because the 21 possible two-factor interaction patterns for the seven factors could be calculated by the experimenter, it would be possible to compare the column in which the interaction effect was detected to those patterns. This would allow the experimenter to narrow the likely contributing interaction to one of two or three. Depending on knowledge of the process from theory or past experience, it might be possible to choose one of the possible interactions as being the most probable.

In this case, a rubber technologist would immediately consider that the accelerator might be activating the sulfur so much as to have major effects on cross-link density, so the sulfur-accelerator interaction could be the cause.

Examination of column five shows that it does have the pattern of the interaction of columns four and six, which are sulfur and accelerator.

At this point, another technique can be considered, which can help pin down the source of interactions. If a particular interaction is strongly suspected, let's say the AB interaction, the design can be altered in a very simple way and the experiments run again, with a very high probability of confirming or denying the suspected interaction.

The alteration is to merely reverse all the signs for either of the factors involved, in this case, either A or B. Whenever the sign of a single factor is reversed in the experimental array, the sign of any interaction to which it contributes will also automatically change. This means the experimenter can repeat a pattern of experiments except that, for the A factor, every plus must become a minus and vice versa. If the observed effect in the column to which the AB interaction fell then reversed sign, the conclusion could be drawn that it was an interaction being observed instead of the main factor effect for that column.

In this case, the final design was a repeat of the reflection but with column five, sulfur, reversed in sign as shown in Fig. 10.4.

In Fig. 10.4, column five reverts back to a positive number from its previous negative level, confirming that it is an interaction of sulfur that is appearing there instead of just the curative contribution.

One side benefit of using folded-over designs is that the two halves can serve as blocks that demonstrate the effect of running the two series at

Reflection 2	Polymer	Black	Clay	Sulfur	Cure	Accel	PrsAid	Tensile Strength
Run 17	+	+	+	+	−	−	+	2,328
18	+	+	−	+	+	+	−	2,419
19	+	−	+	−	−	+	−	2,033
20	+	−	−	−	+	−	+	2,065
21	−	+	+	−	+	−	−	2,415
22	−	+	−	−	−	+	+	2,265
23	−	−	+	+	+	+	+	2,128
24	−	−	−	+	−	−	−	1,991
− Total	8,800	8,217	8,740	8,779	8,617	8,799	8,859	17,644
+ Total	8,843	9,426	8,903	8,865	9,026	8,844	8,784	
Difference	43	1,209	163	86	408	45	−75	
Effect	10.8	302.2	40.8	21.5	102.0	11.3	−18.7	

Figure 10.4 Variation of design of Fig. 10.2 but with column 5 reflected

different times. The grand averages of each block should fall within the limits of normal variation, because it is the same process being run with complementary designs. If there is a significant difference between the two averages, it may be an indicator of how much the process can change as a result of being run at different times (or different lots of raw materials, etc.).

However, if proper consideration can be given to possible interactions before the first experimental series is drawn up, it is often possible to avoid any need for the use of folded-over designs. Selection of designs fitted to expected interactions can eliminate the need for such follow-up work. For instance, up to four factors that well may interact can be combined with four noninteractive factors in a 16-run design (see Chapter 9).

In summary, a folded-over design is a convenient and efficient way to sort out main factor effects from interactions once a fractional factorial design has produced data that show definite effects, some of which are suspected to originate from interactions.

11 Nomenclature and Design Variations

Occasionally, students of design of experiments (DOX) become confused by the different conventions used in various books and commercial software. Although any single design pattern is actually constant in its content and capability for use, differing ways of denoting factor settings and rearrangements of the patterns can make two identical sets of experiments appear quite different. One such example is the contrast between the commonly taught Taguchi L8 pattern and the expanded classic eight-run design.

The classic design starts, as shown in earlier sections, with all the combinations possible for three factors at two levels, with the coded levels designated as − and + (Fig. 11.1). (For three-level experiments, this system merely adds a 0 level, which is midway between the − and + levels.)

Some experimenters favor the convention in which the − levels are equal to whatever is considered standard for the process, and the + levels are some change from the standard. Thus, if the process had always been run at room temperature and the experiment was to explore a lower temperature, the minus level could be 20 °C and the plus level would be 10 °C; the standard catalyst is minus and the new alternative is plus.

Others prefer to use intuitively understandable relationships, so that the plus level is always the higher on whatever scale of variable is to be used;

Run 1	+	+	+
2	+	+	−
3	+	−	+
4	+	−	−
5	−	+	+
6	−	+	−
7	−	−	+
8	−	−	−

Figure 11.1 Classic 3 × 2 full-factorial design

in the case of the old and new catalyst, most experimenters do tend to assign the minus designation to whatever is considered the known standard. However, as long as the individual running the experiments and interpreting their results knows what convention has been assigned, it really makes no difference whatsoever which it is.

To produce a saturated eight-run design, four more columns of factor settings are added, as was shown in Chapter 3. Each column has to have four minus and four plus settings in it to keep the design balanced in general, but to make it orthogonal, the order of the minus and plus levels is derived from the first three columns by multiplying them in pairs (AB, AC, and BC) to produce three new columns and then as the product of all three (ABC) to produce the last column. The resulting matrix is seen in Fig. 11.2, again in one of its most basic forms. (Some prefer to show it with the runs reversed, so that run 8 below becomes run 1, etc., whereas others would have the order of column C exchanged with that of column A. These are just conventions among which anyone may choose as they like.)

However, there is nothing sacred about the pattern as such, so that if the rows are randomized (as they might be for actual use) and the columns are moved around for some reason (to list the factors alphabetically, perhaps), nothing really changes. Thus the matrix in Fig. 11.3 is actually identical to the one in Fig. 11.2, even though it may not be immediately recognizable.

The run-order pattern contained in the original matrix is very often not the order in which the actual runs are made, because randomization is often used in implementation of experimental designs. No single way of distinguishing the original order from the randomized order is universally accepted, but contrasting terms such as "standard versus random order," "experimental versus laboratory order," and "design order versus order of running" are used. The use of only one convention in any lab is essential.

	A	B	C	D	E	F	G
Run 1	+	+	+	+	+	+	+
2	+	+	−	+	−	−	−
3	+	−	+	−	+	−	−
4	+	−	−	−	−	+	+
5	−	+	+	−	−	+	−
6	−	+	−	−	+	−	+
7	−	−	+	+	−	−	+
8	−	−	−	+	+	+	−

Figure 11.2 Saturated seven-factor, eight-run design

	E	C	G	D	B	F	A
6	+	−	+	−	+	−	−
2	−	−	−	+	+	−	+
8	+	−	−	+	−	+	−
4	−	−	+	−	−	+	+
5	−	+	−	−	+	+	−
3	+	+	−	−	−	−	+
7	−	+	+	+	−	−	−
1	+	+	+	+	+	+	+

Figure 11.3 A rearrangement of pattern of Fig. 11.2

	E	C	G	D	B	F	A
Run 6	−	+	−	+	−	+	+
2	+	+	+	−	−	+	−
8	−	+	+	−	+	−	+
4	+	+	−	+	+	−	−
5	+	−	+	+	−	−	+
3	−	−	+	+	+	+	−
7	+	−	−	−	+	+	+
1	−	−	−	−	−	−	−

Figure 11.4 Reflected pattern of Fig. 11.3

Just to add further to the possible variations, the reflection of designs is one more common confusing factor. For instance, the design in Fig. 11.4 is a reflection of the one shown in Fig. 11.3 and is therefore also equivalent to the basic eight-run from which Fig. 11.3 was derived, but the resemblance is now even more remote. (Notice that there is now one run with all the settings at minus, whereas previously there was no such run but instead there was one with all the settings at plus.)

Reflected designs contain the same amount of information as their originals, but they have to be analyzed slightly differently. The main point is to know that interactions now appear changed in sign; multiplying the A settings by the B settings results in a column of minus and plus signs which is not matched by any of the column patterns shown. However, one of the other columns will have a pattern that is a mirror image of the AB pattern. This is illustrated in Fig. 11.5 by showing the A, B, and D columns from the matrix in Fig. 11.4 plus an AB column calculated as the product of A and B.

Note in Fig. 11.5 that the D column is a mirror image of the AB interaction column. This means that the contrast shown by analyzing the responses using the D column pattern of addition and subtraction is the effect of the AB interaction but reversed in sign.

The estimation of two-factor interaction effects in reflected designs is always reversed in sign (but main factor effects are not). This is important background for the use of folded-over designs, which were discussed in Chapter 10.

Plackett-Burman (P-B) geometric designs also fit the pattern of the basic saturated fractional factorials, and the P-B 8 shown in Fig. 11.6 is equivalent to all the eight-run patterns already shown.

Now that most of the possibilities of altering the appearance of design patterns using conventional notation have been covered, Taguchi's particular conventions and patterns can be shown. Dr. Taguchi prefers a system in which levels are designated as 1 and 2 (and in three-level designs, they become 1, 2, and 3). His original L8 pattern is shown in Fig. 11.7, followed by a rearranged version (Fig. 11.8) that has the columns in the same order as the

	A	B	AB	D
Run 6	+	−	−	+
2	−	−	+	−
8	+	+	+	−
4	−	+	−	+
5	+	−	−	+
3	−	+	−	+
7	+	+	+	−
1	−	−	+	−

Figure 11.5 Selected columns from Fig. 11.4 plus an AB interaction column

Run 1	+	+	+	−	+	−	−
2	+	+	−	+	−	−	+
3	+	−	+	−	−	+	+
4	−	+	−	−	+	+	+
5	+	−	−	+	+	+	−
6	−	−	+	+	+	−	+
7	−	+	+	+	−	+	−
8	−	−	−	−	−	−	−

Figure 11.6 Eight-run Plackett-Burman design

Run							
Run 1	2	2	2	2	2	2	2
2	2	2	2	1	1	1	1
3	2	1	1	2	2	1	1
4	2	1	1	1	1	2	2
5	1	2	1	2	1	2	1
6	1	2	1	1	2	1	2
7	1	1	2	2	1	1	2
8	1	1	2	1	2	2	1

Figure 11.7 Original Taguchi L8

Run							
Run 1	1	1	1	2	2	2	1
2	1	1	2	2	1	1	2
3	1	2	1	1	2	1	2
4	1	2	2	1	1	2	1
5	2	1	1	1	1	2	2
6	2	1	2	1	2	1	1
7	2	2	1	2	1	1	1
8	2	2	2	2	2	2	2

Figure 11.8 Rearranged Taguchi L8

basic eight-run design but in which the runs have been reversed in order from the basic eight-run pattern.

Using the 1, 2 convention instead of the −, + convention of course makes no difference at all in the meaning or use of the matrix but is merely a matter of choice. Most of the time when the patterns are converted from one to another, the 1 becomes − and the 2 becomes +, but, like all the conventions that also can be reversed, with no change to the content of the experimental pattern. (However, it may mean the interaction columns reverse in sign as reflected designs always do.)

In this section, the comparison and contrast of differing nomenclature systems have been explored so as to help raise understanding of how designs relate to each other and avoid confusion when different literature or software packages are examined. When questions arise about any particular design, it is always possible to convert all numbers or alternate codes to the −, + convention and then work through interaction relationships in the data by comparing the product of any two columns to the sequence in every other individual column (or, in the possible case of a reflected design, to the mirror image of every other individual column).

12 Estimation of Scatter

The very convenient scree plot often serves quite well in enabling an experimenter to tell the difference between minor variation in test data and significant effects resulting from deliberate changes in control factors. However, there are three other methods for separating real effects from experimental scatter. They involve the use of methods requiring a little more knowledge of Statistics, and some calculations have to be made; some readers may find this chapter to be more than they need to use designed experiments, but reading through it once for background is probably worth the effort.

All three methods involve computing confidence intervals around each difference between averages from contrasting runs. The confidence intervals will vary in width depending on some estimate of the level of random error in the system. There are three possible sources by which the random error can be estimated:

- Historical data,
- effects in "empty" columns, and
- effects from repeated runs at the same conditions.

For instance, in the example in Chapter 8 using the minimum viscosity of a rubber compound (ML), long history of use of the test instrument indicated that the total scatter in readings when the same batch is retested numerous times is about ±1 unit. (Thus the standard deviation of the test would be about ±0.5 unit, because four standard deviations make up the 95% confidence interval.) Therefore, for this property, the confidence interval of any effect greater than +1 or less than −1 will not overlap 0, so the differential found between the two group averages from the experiment (in this example, for instance, the group of four runs which used the low-viscosity polymer and the group which did not) can be accepted as having demonstrated a real contrast in the response caused by the factor being evaluated. This is because that differential between those two averages has been shown to be significantly different from 0, that is, different from no real difference. This is illustrated in Fig. 12.1 for the ML data.

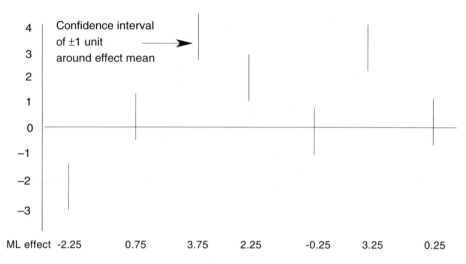

Figure 12.1 Confidence interval chart for ML effects

Note in Fig. 12.1 how three of the effect intervals do overlap the zero line, indicating the measurements are more likely to represent normal system error than any change in the response actually caused by the particular control factor. Of course, the scree chart for this property shows the same three factor effects to be close to zero and each other, which leads to the experimenter making the same judgment about their significance.

Unfortunately, it is often the case that good historical data are not available, and it is necessary to have within the experiment itself some source of information on its scatter level. A common method of arranging this is to leave one or more columns vacant of main effects and probable interactions and to use the observed effects as indicators of system error.

In the example in Fig. 12.1, if the second, fifth, and seventh columns had been unassigned and interactions were considered to be minor or nonexistent, those three effects (0.75, −0.25, and 0.25) could be used to estimate the standard error of the test. This would be done by following the principle that the pooled standard error is equal to the square root of the total of the squares of each unassigned factor effect divided by the number of such effects. Stated mathematically, this is

$$\text{Standard error} = \sqrt{\frac{(0.75)^2 + (-0.25)^2 + (0.25)^2}{3}}$$

$$= \sqrt{(0.5625 + 0.0625 + 0.0625)/3}$$

$$= \sqrt{(0.6875)/3} = \sqrt{0.229} = 0.478$$

Interestingly enough, this is very close to the 0.5 value derived from prior experience and would lead to the same conclusions as to the validity of the other three effects.

Strictly speaking, this estimate of the standard error is known to be inaccurate, because any such estimates based on very limited sample sizes (certainly <20) definitely tend to be too small. Because the sample sizes are too small, an interval must be wider than ±2 estimated standard errors to really contain 95% of the possible distribution being examined.

How much wider the intervals need to be is a question that was answered in the early part of this century by a mathematician working in quality control for the Guinness brewery in Ireland. He derived a particular set of distributions, called the t-distributions, in which the individual t elements represent the number of standard errors needed to produce accurate confidence intervals for samples with N degrees of freedom. Selected sections of these distributions appear in tables, broken down by the confidence interval desired (often 90, 95, and 99%). A small subsection of one such table is shown in Table 12.1.

Because the standard error calculated from the example above was based on three empty columns, the degrees of freedom (df) would be 3, and 3.18 is the t number from the table for a 95% interval and three df. This means that the more accurate equation for a 95% confidence interval would be ±3.18 ✤ × 0.478 units, or three units in total width instead of the two-unit width used as a first approximation. (In this particular example, nothing would have changed in the conclusions about the meaning of the effects, but in other cases

Table 12.1 t-Distribution Values for Two-Sided Confidence Intervals

	Confidence Level, %		
df	99	95	90
1	63.70	12.70	6.31
2	9.92	4.30	2.92
3	5.54	3.18	2.35
4	4.60	2.78	2.13
5	4.03	2.57	2.01
6	3.71	2.41	1.94
7	3.50	2.36	1.89
8	3.36	2.31	1.86
9	3.25	2.26	1.83
10	3.17	2.23	1.81

df, degrees of freedom

there could be different interpretations of the data if corrected confidence intervals were used.)

As the df term gets larger, the t numbers become smaller, and in fact at 10 df, the t number is 2.23, only about 10% larger than the uncorrected level of two standard errors that applies in theory to very large sample sizes.

Conservative statisticians would disallow pooling the effects of columns two, five, and seven in the real experiment, because two of them did have factors assigned to them. Only the last column was actually empty, so its squared effect (0.0625) would then have to be multiplied by 12.7 (95% t-distribution for 1 df) to give an estimate of one-half the confidence interval width at about 0.8 units; the 1.6-unit full width of the interval in this case happens to be less than those from either historical data or the pooled estimate.

Use of the t-distribution is predicated on underlying assumptions, which include not only normality and independence of the parent distributions of test results but also that their variances are equivalent. As sample sizes become smaller, the sensitivity of the procedure to deviation from these assumptions becomes quite high so that use of sample <10 is risky, and <3 is very risky indeed.

One of the major characteristics of the Taguchi approach is to disregard the requirement for using only effect levels from truly empty columns to estimate system error. Dr. Taguchi calls for taking all effects that are clearly lower in magnitude than the uppermost effects and using them in a pooled estimate of error. This does make possible more effective use of the data to evaluate system error and provide quantitative estimates of the value of observed effects.

The other classic method for estimating scatter is to use replicates of one or more sets of experimental conditions. The pooled standard deviation of test results from those repeated experiments is perhaps the best estimator of all for the system error. Assuming that only one condition is replicated, the standard deviation of that subgroup is used in the same equation seen earlier to estimate the standard error for the difference of two averages,

$$\text{Standard error} = 2s/\sqrt{n}$$

except n is the number of all the runs in the experiment, not just the number of replicates. However, the df for the t-distribution used to calculate the confidence interval is the number of replicates minus 1.

Much of the time, replication is used in more advanced experimental designs that are typically analyzed using analysis of variance through a computer software package, so the details of that calculation will not be provided here.

In this section we saw:

- Three different methods for estimating what is normal scatter in a series of experimental runs,
- that the use of very limited amounts of data in making those estimates is highly risky,
- an example of the contrast between classical techniques and the Taguchi approach, and
- that simple replication of runs is recommended when possible as the best method for determining scatter levels in an experiment.

13   Sizing of Experiments

The usual approach to an experimental design, as mentioned in Chapter 6, is to select the control factors of special interest and the number of levels at which they need to be evaluated and then to choose a design that allows exploration of that combination of factors and levels most efficiently. The theme of doing the least possible experimental work to yield needed data is a principal one in most approaches to design of experiments (DOX). However, finding the design that contains the fewest number of experimental runs containing the factors and levels chosen is not the only way to judge an experimental strategy as being appropriate. As in Chapter 12, some reference to statistical procedures will be used which may or may not be of particular interest to different readers.

A valid comparison of data has to be able to find real differences of whatever extent is important to the person doing the comparing. Suppose someone were trying to measure metal ball bearings of a quarter-inch diameter and they had three tools with which to measure diameter. One tool is a wooden ruler, one is a standard micrometer measuring out to 0.001", and one is a laser gauge precise to millionths of an inch.

Using a wooden ruler to check the diameter of a small metal ball is hard to do with any real precision. Many people with good eyesight could tell the difference between two bearings using the ruler when one is 0.250" and the other is 0.325", but if the second were only 0.260", it would be almost impossible to see the difference of 0.01" on the scale of the ruler. In the terminology of quality control, the ruler is not able to discriminate differences of that size.

However, the micrometer, which could measure to the nearest 0.001", would easily discriminate a difference of 0.01", so if the product specification called for bearings to be between 0.247" and 0.253" (which is the same as 0.250 ± 0.003"), using the micrometer would be appropriate. If the bearings were for some super special usage and the specification was 0.250 ± 0.0005" instead, then it would be necessary to use the laser gauge, because the specification calls for a higher degree of discrimination than can be attained using the micrometer.

Test data from experiments can also be evaluated in terms of capability to discriminate, which affects how many experiments need to be run. The discrimination capability of an experimental design increases with the number of runs in the design.

A prime characteristic of DOX is that comparisons are made between averages of some response instead of between single data points. The Central Limit Theorem (CLT) reveals that there are substantial advantages to using averages instead of point data. The question then arises (as usual) as to just how great a difference between a pair of averages there needs to be to generate real confidence in the meaning of that difference.

Answering that question begins with the use of another statistical concept, that of the variance of the difference between the averages. Variance (V) is a number calculated from a distribution by statisticians to characterize the width of the distribution. Most of the time, ordinary people use the standard deviation (SD) for that purpose and, in fact, the standard deviation is derived directly from the variance. (The standard deviation is equal to the square root of the variance.)

It is not essential to go over the equations defining the variance. What is important is that the variance of the difference between two averages (V_D) is equal to the weighted sum of the variances of each group, which is expressed in a comparatively simple equation,

$$V = (V_1/n_1 + V_2/n_2)$$

and if the group variances are equal (which is likely when they're all from the same process), then $V_1 = V_2 = V_x$ and the equation becomes

$$V_\Delta = V_x(1/n_1 + 1/n_2).$$

The number n_1 is the total number of experimental runs used to calculate one of the average responses, and n_2 is the total number of runs for the other average response. If, as is usually arranged in designed experiments, the test groups are of the same size, then $n_1 = n_2 = n/2$ (where n is the grand total of all the experimental runs), and the next transformation is to

$$V_\Delta = 4V_x/n.$$

Recall that the standard error (SE) is the same as the standard deviation of the test groups, which is equal to the square root of the variance V_Δ, so taking the square root of each side of the equation leads to

$$SE = 2\sigma_x/\sqrt{n}$$

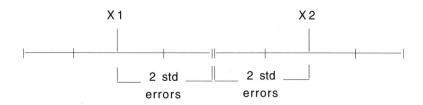

X1 and X2 are the two group averages, each shown in th center of its confidence interval

Figure 13.1 Nonoverlapping confidence intervals for group averages X_1 and X_2

where σ_x is the standard deviation of the data in the groups.

One use of the standard error is to construct confidence intervals, as seen in earlier chapters, with such intervals being the range of numbers within which some true number can be found at the stated level of confidence. In the case of the difference between two averages, ±2 standard errors covers roughly 95% of the possible spread of the true difference.

To ensure that the difference is real, the interval size is doubled so that each of the compared averages has at least a 95% chance of not falling within the confidence interval of the other. This means that if four standard intervals separate the two group averages, there is good confidence (approaching 95%) that they are different. This is demonstrated in graphic form in Fig. 13.1.

Because one standard error is equal to $2\sigma_x/\sqrt{n}$, four standard errors are equal to the quantity $8\sigma_x/\sqrt{n}$. This is the amount of difference needed between two averages to be about 95% sure they are different. This relationship can be expressed in the equation

$$\Delta = 8\sigma_x/\sqrt{n}$$

where Δ is the difference between the two averages.

This can be rearranged to

$$n = [8/(\Delta/\sigma_x)]^2.$$

The Δ/σ_x term is the ratio of the observed difference to the scatter in the data, which is another form of the signal-to-noise ratio.

The standard deviation of the sample distribution, σ_x, is unknown, and usually the only available estimator for it is s, the standard deviation of the sample. The substitution is made and the equation becomes

$$n = \cancel{[64/(\Delta/s)]^2} \quad 64 : (4/3)^2$$

where n is the number of data points (experimental runs) needed to find a real difference when the signal-to-noise ratio is Δ/s. (If any better estimate of σ_x

is known, for instance from good historical data on the process under investigation, it can be used instead of s.)

More detailed derivations of this equation indicate that it is slightly on the conservative side, i.e., the number 64 may be larger than needed to allow good discrimination between the averages being compared. Some statisticians find that 49 may be used instead; in practice, the calculated n is seldom an even whole number, so typically it is rounded up or down to some convenient size for actual use.

Use of the equation proceeds as follows: a salesperson proposes the use of a new additive to increase the yield point of a particular plastic, which has been running around 50% elongation (or 0.5 strain level). A possible increase to 0.6 strain is mentioned, which would provide welcome improvement for the material.

In this case, Δ is 0.1, the amount of strain level increase anticipated. If the standard deviation for yield strain determinations happened to be about 0.1, the signal-to-noise ratio would be 1 and the equation would indicate that 50 or more data points would be needed in the comparison to reliably find the hoped-for improvement. (This would be 25 control batches and 25 batches with the new additive.)

If the standard deviation of the yield data were 0.05 instead, the ratio would become 2, the squared term would be 4, and between 12 and 16 batches could be used to demonstrate the effectiveness of the additive. If the anticipated increase in yield strain were 0.2 and the standard deviation was still 0.05 so that the ratio was 4, the squared term would increase to 16, and only four batches of material could show whether or not the additive performed as promised.

In planning out a DOX strategy for only three factors at two levels, at first glance the use of the basic eight-run design seems fully appropriate, and in many cases it is, but an eight-run series is not capable of finding effects if they are not at least 2.6 times as great in extent as the basic property scatter from the process. Thus, if a 10% increase in reaction rate through the use of some combination of catalyst concentration, temperature, and pressure was being explored, but the standard deviation of rates was 5%, it might necessary to use a 16-run design or replicate the entire eight-run design once to get sufficient data to understand the process.

In some cases, the need to have a higher total number of runs might lead to consideration of different designs altogether. For instance, in the situation above, instead of replicating an eight-run, two-level design once to get to a 16-run total, a three-level design using 17 runs might be used to get both the needed amount of information and a more advanced level of information.

More typically, the tactic used is to improve the signal-to-noise ratio by either decreasing s (possibly through improved testing methods) or increasing Δ, the amount of the effect. The latter is most often attempted by using a greater range of the control factor level so that the difference between the two property averages is inflated. Instead of varying temperature from 190 °F to 210 °F to check the effect on reaction rate, it might be changed between levels of 175 °F and 225 °F, because a 50 °F differential is much more likely to show a contrast. This is called using bold limits for the factors.

This chapter's main contribution is that, in drawing up experimental designs, consideration must be given to the comparison of the anticipated effect and the background scatter of the process. The equation $n = [64/(\Delta/s)]^2$ is a very useful guideline in helping to ensure that adequate data are generated.

$$n = 64 \div (\Delta/s)^2$$

14 Development of an Experimental Strategy

In both industrial and laboratory situations, a wide variety of challenges can and will be encountered. They may range from starting up the new machine, which has numerous unfamiliar controls, none of which is well understood, to optimizing a process that has a long history and only a few well-known major factors affecting it. The broad sweep of possible problems to be investigated makes it impossible to lay out a simple cookbook approach that is universally applicable. Instead, it is necessary for the scientist/engineer/ technologist to be flexible in evaluating and attacking each new situation.

The military analogy is that of using strategy (long-term planning) and tactics (immediately applicable method). It is very seldom that the experimenter can run a single design series and achieve thereby understanding of a process so complete that no further explorational, optimizational, or confirmational work need be done to justify a high level of confidence that everything important about the process is known. Therefore, when first examining the process, the experimenter will need to begin deciding what strategy will be used for the investigation and then which tactics will be used to implement that strategy.

Complex and novel processes often require a series of experimental designs for full understanding, and it is usually the case that only the first or second step of several can be planned out in detail; the structure of the following steps will depend on what is learned initially. The strategy for analyzing the process may remain unchanged, but the tactics, i.e., the particular designs, factor combination, or factor levels used may need to be changed as the investigator digests the first sets of data. Flexibility and adaptability are key characteristics of every experimenter.

When the process is very new or unfamiliar and the first level examination of it reveals a substantial number of possible control factors (6–20), the use of two-level screening designs becomes almost mandatory. Their purpose is to clearly resolve which factors are truly significant to the process so that the next stage of experimentation can be laid out most effectively.

One major question becomes, what particular range will the two settings cover for each experimental factor? In the ideal situation, the range is chosen to be on the bold side, that is, wide enough so that if no effect can be found for the factor across the range, that factor can then be dismissed from further consideration with a high level of confidence. (Of course, too bold a range has its own dangers. A range of 40 °C for some chemical reaction is very likely to show whether temperature has an effect, but a range of 200 °C may generate data that are misleading or useless.)

Depending on what is known of the process, the screening designs may be geometric or nongeometric. When interactions are strongly believed to be unlikely, the geometric designs (8- or 16-run most often) can be used, and if interactions are a concern, the nongeometric 12- and 20-run designs become appropriate.

Once the initial screen has been run, it should become possible to determine which factors are the most important and whether the ranges used in the experiment were reasonably appropriate. For instance, if temperature had been a factor and was varied from 100 to 200 °C and all the runs at 100 °C failed to react at all but those at 200 °C did react in a moderate degree, the conclusion could be that the range should have been from 150 to 200 °F or perhaps from 200 to 250 °F. In this case, the work may be salvaged in the sense that the half of the design using the 100 °F settings can be run again but at the 150 or 250 °F setting to result in the design now thought to be more valid.

The question of breaking the experiment into two separate halves done at different times, which amounts to confounding temperature with experimental sequence, then comes up and may have to be dealt with by limited replication, but that is still less work than repeating the entire experiment. The idea of building a new experimental pattern using data from a previous one is a useful DOX tactic. Many instructors recommend the use of a series of three or four factor designs, in which each has one surface in common with a prior design, for logically exploring a process.

For example, if a process has four possible control factors, the first design might explore A, B, and C at −1 and +1 settings, with D fixed at a −1 level. If the data showed that the −1 level chosen for C was actually too low, only four more runs could be made with an alternate level of −1 for C, which would allow building a new experiment off the old. This is illustrated in Fig. 14.1, with the first eight-run experiment drawn as the traditional cube form, with the settings for each factor associated with each of the eight corners and each surface assigned to a −1 or +1 level for one of the factors.

Once the −1 level for C is determined to be inappropriate, a "new" side is generated with the "−1a" level of C, and the juxtaposition of that side with the

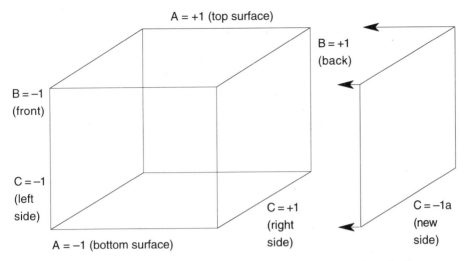

Figure 14.1 Possible changes in an eight-run design shown in cube form

existing +1 C side produces a new cube form. This is merely a graphic representation of how the experimental data from the original eight and secondary four are used to make a fresh set of eight data points for analysis.

Another way to use this method would be that, if in the first experiment it were shown that the C factor actually had no effect on the process, then four runs with the D factor at a +1 setting could be combined with either C side to make a pattern with A, B, and D as factors. This stepwise approach can be an efficient way to explore a process.

With the assumption that, finally, the data from the screening design indicate which factors are the most important ones to the process, the next step might be to draw up another two-level design for those factors which need to be further explored. (Not all factors found to be important might have to be part of the next stages of work; for instance, if the screening design showed one catalyst to be much better than another, the choice of catalyst would then become fixed for future experiments.)

However, this second design would be one that would allow specific investigation of interaction terms. It might also use somewhat adjusted levels of variation of the factors if the first experiment indicated to the investigator that the beginning choice of factor levels could be improved. In that case, comparatively few of the data points from the first experiment might be used again in the second, and in fact an entirely new pattern might be used.

However, if no adjustment of factor levels were to be made, the second experiment might be just the reflection of the first, because that is one simple way to look for the presence of interactions in a process. This kind of choice is what the experimenter faces routinely in the course of such work.

Once the second series has been performed and any important interactions revealed, the experimenter is then able to make some judgments on the possible model for the process and decide what the next step should be. It might be a simple expansion of the second experiment to just take the levels of the factors further toward the direction of better results, or it may be a multilevel response surface design to explore in detail process optimization within the experimental region where the optimum is now believed most likely to be found.

Finally, in the fourth or fifth step, the experimenter selects a narrow experimental region that he or she believes the data show to be the most probable locale for an optimum process. Two or three runs at identical conditions might be performed to confirm the process output and level of scatter there, or a small design (typically a four-run) covering the narrow area could be used instead to pin down the optimal factor settings. This series of experiments would fall into a flow diagram similar to the one in Fig. 14.2.

Not every investigation will have to encompass three to five experimental series. Sometimes there are only a few factors known to be important to the process, and the first design used can include interaction terms. Perhaps the data will then quickly point to the better combination of factors to improve the process, and one more experiment can bring the knowledge of the process to the needed level of detail.

On other occasions, only a few factors might be involved, and the question of what levels to use for them might be considered particularly

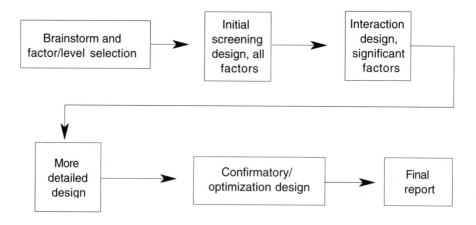

Figure 14.2 Simplified flow diagram for a project using designed experiments

urgent (or, if they represent discrete variables, which of several is the best candidate for use.) In such cases, the use of the Taguchi L9 or L18 designs to immediately explore broader ranges of factor settings might be very effective. Again, these are routine considerations for the experimenter to encounter, and as experience with DOX is gained, so will the person's ability to react and make good choices in experimental strategy and tactics.

The experimental scientist looking at a new technical challenge is like a mechanic approaching a poorly functioning car motor. The mechanic brings with him or her a toolbox filled with numerous wrenches, screwdrivers, socket sets, etc., some of which are general purpose tools and some of which are built for very specific tasks. The scientific investigator has available full factorials, half factorials, many two-level screening designs, some three-level screening designs, assorted response surface designs, and perhaps others such as Latin squares and mixture designs.

The mechanic will size up the problem and probably choose one of the general purpose tools to begin engine disassembly and diagnosis. Depending on what is found, more specific tools and techniques are used to continue the engine disassembly and analysis of the problem. Ultimately, an understanding of the problem is reached and the appropriate repair is made, but the mechanic did not know when first looking at the engine how to predict every step of the work.

In similar fashion, the scientist does a first-level inquiry into a process, probably using a general purpose screening design, and then, as data provide helpful inputs, he or she moves on thoughtfully into selection and use of other designs until an understanding of the process is gained. (A more advanced practice is to progress to development of not just the ideally most productive process but further to the best combination of productivity and consistency for the process; this is referred to as a robust process, but expansion of this concept will be deferred for now.)

The summary of this chapter is quite basic. The broad assortment of problems to be dealt with by the scientist or engineer will sooner or later pose difficulties, not all of which can be met with a single approach in experimental techniques. Therefore, some knowledge of what differing techniques, or which succession of techniques, can be applied to different situations is of critical value to the investigator. The trap of thinking that DOX will always allow complete solutions to be found very rapidly, i.e., one set of experimental runs, for any problem should be avoided. A realistic point of view is that any problem of complexity is almost certain to require a campaign of at least some length to be fully solved.

Continued referral to literature and the experience of others while gaining personal experience will always aid the individual in becoming an ever more capable experimenter. User-friendly software can be of great help and convenience to the experimenter, but there is still no substitute for a thinking and adaptive human intellect in investigating and understanding any phenomenon of interest.

Critical themes in this chapter include:

- Problems of any complexity are not often understood and resolved by means of a single designed experiment;
- when facing complex problems, it is frequently necessary to think out both the long-term approach (strategy) and the immediate steps (tactics, which means individual designs);
- designs that sort out groups of discrete factors (L18), do basic exploration of process variables and simple interactions (geometric two-level designs), check out main effects only (nongeometric designs), or evaluate nonlinear effects (L9) are among the tools available to the experimenter; and
- tactics can include a series of similar designs that build off each other, a sequence of screening, factorial, and response surface designs, or whatever sequence the data lead to in the progression of designs.

15 Basics of Response Surface Methods

It would be very convenient if all responses to changes in important factor levels were simple and linear for any process of interest, at least within the range of factor levels easily controlled or the range of response desired. Tuning a process to some desired output would then be easy.

However, things just don't work that way all the time. Some more advanced techniques have to be used to understand more complex processes. These techniques are called response surface methods (RSM), and this chapter will discuss them in a limited, introductory way. Some mathematical models have to be shown as part of the explanation, but nothing too complicated. All that is strictly necessary for the beginner in design of experiments (DOX) is to read over this and the following chapter to get at least some idea of what RSM is and how it works.

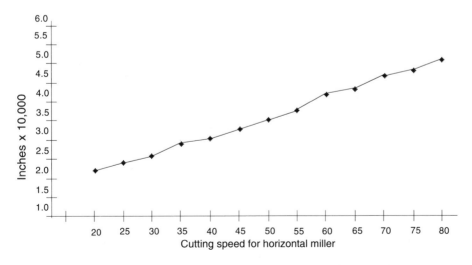

Figure 15.1 Dimensional tolerance vs. machining speed

If a machining process were such that, as the cutting speed changed, the dimensional control of a metal mold varied inversely with the increasing speed, a graphical presentation of the data might appear as in Fig. 15.1.

This type of graph is very informative in itself, but the goal of RSM experiments is to generate a mathematical model, an equation that uses various numbers to describe the process. If the factors examined in the experiments are really the main ones controlling the process, and the data from the experiments are reasonably accurate and precise, then it becomes possible to develop a model that will describe the process validly. This means that using the equation with the actual settings used in the real experiments results in predicted results which match up closely to what was really observed; and this implies that using the same equation with process settings that were not run will still accurately predict the process results under those conditions. The techniques for developing these models are relatively common in higher mathematics but are beyond the scope of this book.

If the data for cutting speed versus part tolerance (in ten-thousandths of an inch) were analyzed mathematically, the resulting model would be a simple linear equation of the type

$$Y = \text{Constant} + (\text{Coefficient} \times X)$$

or in this case,

$$\text{Tolerance} = 1.2 + (0.2 \times \text{Cutting speed}).$$

This is convenient in that, if maximum machining throughput is desired, the supervisor can find out what tolerance is acceptable to the customer and set the miller for the highest speed that can maintain that tolerance.

The constant in the first equation is sometimes referred to as the intercept, because that is where the line will cross the Y-axis if extrapolated. Determining what the intercept and the coefficient are for this one-factor process is the mathematical solution for understanding and ultimately controlling the process.

When three factors are explored for a process, the simplest linear equation is

Linear model

$$Y = \text{Intercept} + C_1X_1 + C_2X_2 + C_3X_3$$

where X_1 is the first factor, X_2 is the second, etc., and each C term is the coefficient for that factor. The use of more factors from the process would just require more terms in the equation.

When interactions contribute to the response, the model must be expanded to include specific terms for those interactions, and for three factors it then becomes

Full interaction model

$$Y = \text{Intercept} + C_1X_1 + C_2X_2 + C_3X_3 + C_{12}X_1X_2 + C_{13}X_1X_3 + C_{23}X_2X_3 + C_{123}X_1X_2X_3$$

The factors and their individual coefficients are identified as in the earlier equation, and the factors also appear the same way in the interaction terms. However, the coefficients for interactions are shown with numerical subscripts that relate to the factors involved in that interaction. For instance, the term $C_{123}X_1X_2X_3$ contains all three factor descriptions, so it is the three-factor interaction and the C_{123} term is the coefficient for that interaction.

These models can be derived from two-level experiments, so any graph of the response can only show a straight line. In many situations this is fully satisfactory, but many natural processes are not truly linear when varied over a range of any real size.

When nonlinearity is anticipated, experiments at more than two levels are necessary for good modeling, because a minimum of three points is needed to describe a response curve. A full nonlinear model for three variables is

Quadratic model

$$Y = \text{Intercept} + C_1X_1 + C_2X_2 + C_3X_3 + C_{12}X_1X_2 + C_{13}X_1X_3 + C_{23}X_2X_3 + C_{11}X_1^2 + C_{22}X_2^2 + C_{33}X_3^2$$

This is called a quadratic model because it includes terms raised to a power, in this case, the squared terms in the last row. This is a full model because it still includes the interaction terms. (Some software will construct models with the squared terms but without the interactions, and which approach is more valid is a matter of occasional dispute.)

Estimating the value of any one coefficient in these equations calls for what statisticians call an independent judgment, which means it requires sufficient information (often referred to as degrees of freedom) to make the analysis; this translates to the design needing to contain at least one experimental run for every term in the equation that might come out of the analysis. The quadratic model above has ten terms in it, counting the intercept, so an absolute minimum of ten runs would have to be made to make any use of a quadratic model with three control factors.

There is no fully balanced RSM design for three factors in just ten runs, but there is one with eleven runs, so that design would have to be used. Most of the time, a larger design would be used anyway, such as a central composite with 15 runs, and the extra runs allow a form of calculation that indicates to some degree how well the model fits the data. "Lack of fit" calculations are common to many computer analyses and will be discussed in Chapter 22, Advanced Topics.

By using the squared terms of quadratic models it becomes possible to describe many kinds of curves that fit the response patterns of various processes. An example of a natural process with highly nonlinear response is the yield of an autocatalytic reaction, such as that seen in Fig. 15.2.

If a two-level experiment was done with time as a variable, using 60 and 90 minutes for −1 and +1, a false picture of the effect of time on the process would emerge. A three-level experiment with 60, 80, and 100 minutes would provide a very different picture and make it possible to develop a more valid model containing higher order terms; this model would allow more accurate prediction of yields at longer reaction times.

In the normal course of process optimization, the necessity of analyzing nonlinear responses will eventually arise. Experiments utilizing more than two levels of factor settings must then be drawn up. The simplest design would be a full factorial at three levels, which for three factors requires 27 runs, for four factors, 81 runs, and for five factors, 243 runs. Obviously, these total numbers of experimental runs would not be attractive to most investigators, and the central question underlying DOX again comes up: How few experimental runs can be made and still yield the bulk of the needed information for understanding the process?

Various statisticians and experimenters have worked out several different but related approaches to answering that question. The classic multilevel designs fall mainly into three groups: central composite designs, face-centered cubic designs, and Box-Behnken designs. (In recent times, other

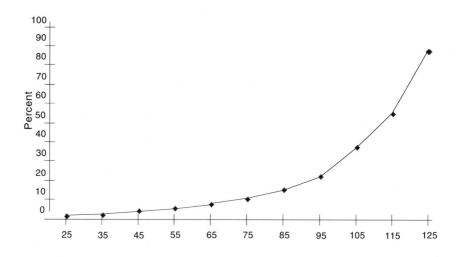

Figure 15.2 Yield vs. reaction time

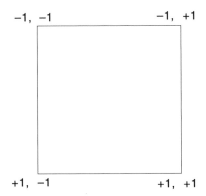

−1, −1   −1, +1

+1, −1   +1, +1

Figure 15.3 Basic 2 × 2 central composite design

types of RSM designs have been developed, such as the optimal families and algorithmic designs, but they will not be discussed here.)

The first of these, and most favored by statisticians, is the central composite. These designs utilize not three but five levels for each factor. A two-dimensional representation of an experimental pattern can be used to illustrate how it works. First is the simple box (Fig. 15.3), which represents the two factors at two levels.

This design can only support linear models, because it only has two levels. The central composite adds first a center point run (designated as having the 0, 0 level) and then outlying points equidistant from the midpoint of each side of the square (Fig. 15.4).

The outlying runs are called star points, because they can be thought of as the tips of a star pattern. For this two-factor experiment, each factor is at five coded levels, −1.4, −1, 0, +1, and +1.4 (which have to be transformed into the real temperatures, pressures, etc., corresponding to the coded numbers). The placement of the star points changes for central composite designs as the number of factors increases.

If temperature was a factor and the original levels chosen to explore the process had been 150 and 250°C (coded as −1 and +1), the five real levels of temperature would be 110, 150, 200, 250, and 290 °C. These temperatures are calculated by subtracting 40% of the range over which −1 to +1 vary (that range is 100 °C in this case) from the −1 level (which is what a −1.4 level means), taking the average of the −1 and +1 levels for the 0 level, and adding 40% of the range to the +1 level to get the +1.4 level.

In the case of two factors, the pattern contains nine runs, which is actually the same size as a full factorial for two factors at three levels. (In fact, if the diamond pattern in Fig. 15.4 is rotated 90°, it will look a lot like a simple

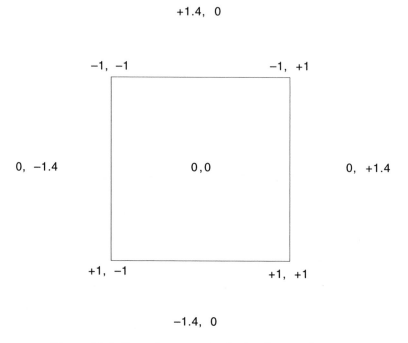

Figure 15.4 Central composite design for two factors

3×3 square pattern.) So, in practice, the two-factor central composite pattern is not usually used.

However, the three-factor central composite uses 15 run conditions (or "treatments" in the language of formal DOX) instead of the 27 needed in a full factorial. For four factors, the total goes up to 25 treatments, for five factors it is only 27 treatments (because of a trick compression of the array), and for six factors the total is 45 treatments.

Sometimes people find the use of the star points inconvenient and instead use the face-centered cubic design. This is like the central composite, but the star points have all been moved back into the confines of the basic pattern, so that it appears as shown in Fig. 15.5.

The analysis of the pattern with three levels of each factor instead of five is not quite as attractive mathematically but is very effective nonetheless. Because these designs are so closely related, the number of treatments used in a face-centered cubic is always the same as in the corresponding central composite.

The Box-Behnken pattern differs notably from the previous two, which basically build onto the original form of a square or cube for two-level

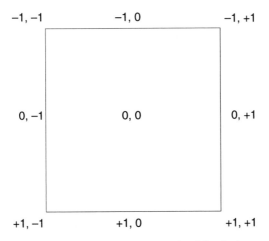

Figure 15.5 Face-centered cubic design

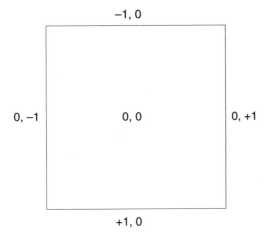

Figure 15.6 Box-Behnken design for two factors

experiments. Instead of using the corner treatments and adding on supplementary points, the Box-Behnken pattern shown in Fig. 15.6 uses only points midway along edges of the square or cube, along with the center point, in its approach.

This difference does not show up well in a two-dimensional display but is easy to imagine for a cube. The standard designs use the eight corners of the cube (think of dice, which have eight corners), whereas the Box-Behnken uses all the points midway along each edge on the cube. There are twelve such points on a cube, which explains how a Box-Behnken for three factors uses thirteen treatments: the twelve edge points plus the center point. For four factors, the Box-Behnken uses 25 treatments, for five it uses 44 treatments, and for six it uses 52 treatments.

None of the figures quoted for these designs include replicated points for the purposes of better estimating scatter. The recommendations by the original developers of the designs call for at least two replications, but more often for four or five replications. It is most common to do all replications using the center point only and to randomly scatter them throughout the entire experimental series order. However, in some designs, the replicates may be of various points outside the center. The data from replicates is used to make an estimate of the scatter basic to the process; in Chapter 20, Full Use of Analysis of Variance, the relationship of estimates of lack of fit and experimental error will be discussed.

In summary, the experimenter has at least three major families of multilevel designs from which to choose when attempting full characterization of a process. The choice of which to use will depend largely on the individual preferences of the investigators and their view of how the particular types and levels of variables to be used will fit best in an experimental array. Analysis of the data can be done by hand if someone is a professional mathematician, but even mathematicians and statisticians nowadays go directly to their personal computers and use an appropriate program for that work. Some explanation of such data analysis is given in Chapter 16.

16 Analysis of Response Surface Method Experiments

Once the data from any of the families of multilevel designs are available, the analysis can proceed. Although it is possible to do the work with pencil and paper, or with a calculator, the computations become quite labor intensive. The curve fitting for nonlinear responses requires regression analysis, a technique not readily familiar to those who are not skilled in calculus and not easy for anyone.

Fortunately, the ready access of computers and appropriate software has made these analyses enormously easier. The first step in using the data is to determine which model best describes the response, whether it be simple linear, interactive, or quadratic. Often a mixed model, which has some linear terms, some interaction terms, and some squared terms, provides the best fit. This is because the actual process is subject to different influences from different factors.

A model with more terms in it will always provide better fit than one with fewer. Fit is measured by a term called R-squared, which is an indication of how much of the variation in the data can be explained by the model being evaluated. However, if a simple linear model had an R-squared of 99% and the quadratic model for the same data had an R-squared of 99.9%, it would not make good sense to use the quadratic model. It gives too little gain in fit for the many terms added to the model.

It should be noted that the R-squared term can be misleading in that, whereas a high R-squared (>95%) is certainly desirable and brings confidence that the model does reflect the reality of how the process works, a lower R-squared (50–75%) does not always mean the model is not useful to understanding the process. Depending on the width of the factor levels, it is possible for the ratio of the change in measured response compared with the background noise of the process to be low enough to keep R-squared small, yet the model can still validly describe the process.

The basic principle behind the choice of a model for any response is to keep it as simple as possible. Each element in the model should be truly contributory, not just another mathematical term that provides a better fit to the response pattern but is not actually based on any real connection with the reality of the process.

Software does not usually make such judgments directly for the experimenter. It will use whatever model for which it is programmed and report the R-squared term. If a linear model had an R-squared of 70% but the quadratic model for the same data had an R-squared of 98%, the experimenter would be motivated to examine the quadratic model closely.

One way of doing that is to scrutinize the regression model term by term. This can be done with paper and pencil, but is done much more conveniently by using the statistical functions normally built into any DOX software.

A statistical analysis of a model will assign a t-number and a corresponding probability to each term in the equation. The probability shown is the likelihood that the assumption of the term's being real will lead to a false conclusion. Thus, when $p = 0.01$ for some term, to a layman it means there is about a 99% chance that the term is really important to understanding the process. Some DOX programs will convert the p number into a confidence statement that the term is real (phrased as the term is being other than 0); in that case, the user sees "99%" instead of "$p = 0.01$."

"High confidence" usually means 95–99+% ($p = 0.01$–0.04) and is considered extremely desirable. If the range is more like 90–95% ($p = 0.05$–0.10), the terms are still very useful. As the confidence estimate drops below 90% toward 80% and lower ($p = 0.10$–0.20), the validity of the term becomes more and more doubtful. With this in mind, the investigator may let the software generate a full quadratic model and then examine it with the intent to trim it down to a simpler yet almost equally well-fitting model.

A software-generated analysis for a quadratic model involving the three factors of temperature, pressure, and time, coded as levels of –1, 0, and +1, might resemble Table 16.1 (although different programs will certainly tend to have contrasts in appearance).

The very first term in Table 16.1, labeled as the constant, is what has also been referred to as the intercept. Its confidence level of 99.9% indicates the equation is in fact validly describing something. With the use of coded levels, the actual number of 3.7 is merely the average level of response measured over all the experimental runs.

The next three terms in Table 16.1 are the main factor effects, and the numbers show directly how much each factor contributes to the response compared with the others. Therefore, temperature has by far the greatest effect, followed by time and pressure. All three terms have confidence levels >90%; in fact, two are >99%, so they clearly are important to the process.

Table 16.1 Typical ANOVA Display from Computer Software Showing Regression Coefficients for Response (R-Squared = 99.7%)

Coefficient	Term	Standard error	t-Value	Confidence coef. $< > 0$, %
3.7132	1 (constant)	0.0612	11.65	99.9
−0.5625	Temp	0.0528	10.65	99.9
0.1007	Pressure	0.0529	1.904	91.4
0.2318	Time	0.0596	3.892	99.6
−0.0214	Temp*Pressure	0.0807	0.954	36.5
−0.3141	Temp*Time	0.0694	2.954	98.5
0.0042	Time*Pressure	0.0112	0.124	18.5
−0.2384	Temp*Temp	0.0448	3.154	99.1
0.0014	Pressure*Pressure	0.0947	0.140	13.7
0.1207	Time*Time	0.0239	1.331	80.5

Confidence figures are based on 5 degrees of freedom (df). ANOVA, analysis of variance.

Table 16.2 ANOVA Display for a Simplified Model Showing Regression Coefficients for Response (R-Squared = 99.2%)

Coefficient	Term	Standard error	t-Value	Confidence coef. $< > 0$, %
3.7132	1 (constant)	0.0612	11.65	99.9
−0.5625	Temp	0.0528	10.65	99.9
0.1007	Pressure	0.0529	2.124	94.7
0.2318	Time	0.0596	3.945	99.7
−0.3141	Temp*Time	0.0694	2.981	98.9
−0.2384	Temp*Temp	0.0448	3.174	99.2
0.1207	Time*Time	0.0239	1.531	86.5

The following three terms in Table 16.1 are the interaction coefficients, and only one of them, temperature*time, appears to be meaningful.

The last terms in Table 16.1 are those useful for describing nonlinearity of the response, and temperature obviously has a nonlinear influence on the process. Pressure does not, and the effect of time is questionable.

The number of degrees of freedom (df) indicates how many runs were made overall; this was a 15-run series with 10 coefficients calculated, so 5 df remain for error.

Due consideration of this table would suggest that three of the terms could be deleted from the model, and with some knowledge of the software being used it is usually possible to tailor-make the model as desired and then rerun the analysis. In this case, the resulting analysis would resemble Table 16.2.

The R-squared term has dropped from 99.7 to 99.2%, which means the elimination of three additional terms from the model only cost 0.5% in goodness of fit. This is very acceptable and demonstrates that those terms did not really hold significance in the model.

Now more df are available for the error estimates, and the confidence levels are adjusted accordingly. The term for nonlinearity with time has risen up to 86% confidence, which gives the investigator more justification for leaving it in the model.

This refined model becomes the basis for prediction of the response throughout the entire experimental volume covered by the settings used. However, extrapolation of the response outside that volume is subject to substantial risk. For instance, if temperature had been varied from 125 to 175 °C in the experiment, prediction of the response at 200 °C might be done with the model derived from the data, but the estimate would carry much less of a confidence factor than estimates made between 125 and 175 °C. (In fact, many DOX software programs will not extrapolate outside the original experimental volume, even though mathematically it is perfectly possible to do so.)

The term "response surface" refers to the concept that actual physical phenomena often have complex patterns of response to changes in several controlling factors. If three factors are involved, it is possible to make a three-dimensional graph with the x-, y-, and z-axes being the levels of each factor. If the model that described the response were a simple linear one, the graph would show the response as a flat plane, with slant and inclination depending on the relative contributions of the three factors.

However, if a complex nonlinear model applies, the response can become a convoluted three-dimensional surface, perhaps resembling a mountain or even a small range of mountains with several different peaks and valleys. This is the response surface for which the investigator desires to generate a reasonably detailed map through experimentation.

When a given process has more than one response of interest, process optimization can become a very complex undertaking. If the factors affect final properties such as the tensile strength, hardness, and fatigue life of the material to be made, plus other characteristics such as cost, yield, and throughput rate, each single response would have its own model. The models might be similar or might cover a range of complexity and be mutually opposed, that is, the conditions to maximize one response might tend to minimize another.

The simultaneous solving of all the individual response equations to predict all properties and characteristics at some setting of the control factors would be a mathematical chore to discourage any but the most dedicated expert. Again, technology comes to the rescue with computers and software, which can perform in a minute or less calculations that in the past would have kept a mathematician occupied for days or longer.

The use of multiple models to simultaneously predict a number of properties for some process is a major part of response surface methodology. Scientists and engineers exploring complicated processes can use a variety of techniques to look for the combination of factor levels that will result in the process being most effective. The definition of what "most effective" means will change from process to process or even from time to time. For one process, maximum throughput might be the greatest need, for another the maximum strength, for another the lowest cost, and so on. The best balance of cost and strength together could be needed for yet another situation.

Many DOX software packages have some form of optimization aids built in, such as multiple plots of responses versus factor levels to help the investigator better visualize how the responses relate to each other. There are even programs that will seek to strike a given balance of properties for the experimenter, using special subroutines and fuzzy logic.

All beginners in the use of RSM have to become familiar first with the types of designs to be used and next with whichever DOX software they will be using in their work. As always, a focused learning effort through actually doing a series of designed experiments will be the best and indeed almost only way to build the skills that allow very profitable use of this methodology, and no amount of knowledge of designs and software can overcome the need for familiarity with the experimental process, good planning, scientific rigor in execution, and common sense in evaluating the data.

Further discussion of RSM analysis falls beyond the scope of this book. Once practitioners of the simpler techniques have gained a reasonable level of direct experience with the basic and some intermediate methods, they will be in a much better position to examine any of the standard texts on DOX. A number of such sources are listed among the references.

17   Basics of Mixture Designs

Some experimenters often or usually deal with developing combinations of ingredients, called formulations, recipes, mixtures, etc., and these combinations always add up to a set total of 100%. One well-known example is simple gunpowder, made up of sulfur, finely powdered charcoal, and saltpeter (potassium nitrate). Other examples might be a recipe for cookies made up of flour, eggs, and butter, or a grease made up of oil, wax, and thickening agent.

In these situations, the additive relationship of the ingredients makes things a bit simpler mathematically. The more common designs have every factor varying independently, which means that, in a three-factor design, knowing what level factors A and B are at in any particular experimental run does not tell anything about where factor C is, so it takes as many pieces of information to describe that run as there are factors in it. However, in a mixture of three ingredients, knowing that A makes up 50% and B makes up 30% of the total immediately indicates that C must be at 20%. It only takes $N - 1$ pieces of information to describe any mixture with N ingredients in it.

This mathematical convenience leads to slightly different ways of looking at the patterns of experiments that can be used, and this part of Design of Experiments (DOX) is called "mixture designs." These designs are used most often by chemists, although on occasion engineers and others can apply them to some types of processes.

One easy way to see how mixture designs differ from factorial designs is to compare how three-factor designs are exhibited in space. A three-dimensional cube has to be used to display the experimental volume of the factorial design, as was shown in Chapter 6. The equivalent-mixture design is displayed by a two-dimensional figure, the simple equilateral triangle shown in Fig. 17.1.

In Fig. 17.1, each of the ♠ points represents a composition of 100% of that ingredient, whereas the ♣ points represent a 50:50 mixture of the two ingredients connected by that edge of the triangle, and the ♥ point is an equal mixture of all three ingredients.

Because of the more efficient mathematical techniques for analyzing mixtures, this seven-run experimental design provides more information on

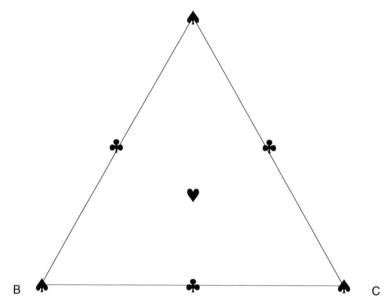

Figure 17.1 Basic simplex design with center point

how the three factors affect this mixture's properties than the eight-run design does for effects of three independent factors on a blend of ingredients. In fact, with this simplex design, a response surface can be constructed for the mixture; that would require a minimum of 11 runs for the 3 independent factors.

Three-factor mixture designs have been used for many years, and special graph paper has often been used to lay out the designs and help analyze the data. An example of a triaxial graph template is provided in Fig. 17.2.

Each line crossing the main triangle in Fig. 17.2 represents a level of use of one of the ingredients. This has been denoted specifically for the ingredient B by the legend at the triangle's base, showing how its concentration goes from 100% at the left-side angle of the triangle to 0% at the right-side angle in 10% increments. Any point on the graph will represent a mixture of A, B, and C that can be read off the sets of concentration lines. For instance, the • in the center of the experimental field is at 33.3% A, 33.3% B, and 33.3% C. (This is called the center point of the design, just as with factorial designs.)

Using the minimal mixture design is quite possible, but just as factorial RSM designs normally have extra design points and replicates to allow better evaluation of the model and estimation of the normal scatter in the process, so do working simplex designs. One regularly used three-factor design adds in three more points evenly spaced around the center point of the basic design, as shown by the ◆ points added in Fig. 17.3.

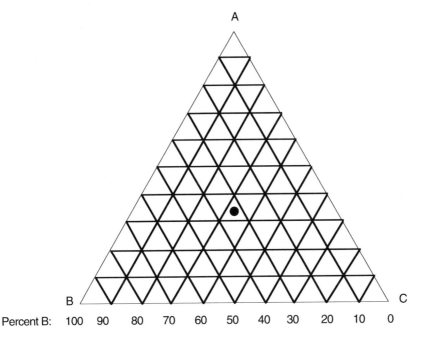

Figure 17.2 Triaxial graph with 10% increments

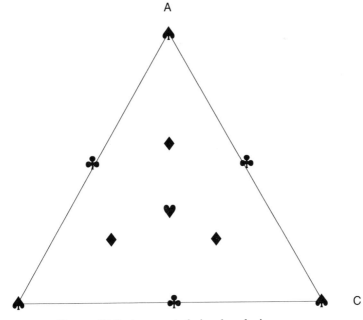

Figure 17.3 Augmented simplex design

The experimental worksheet for the set of runs in an augmented design showing the percentages of the ingredients in each run of the series of mixtures is shown in Fig. 17.4. (Run 7 would be replicated at least twice for a total of 12 runs, and randomization would be applied to the list before the runs were made.)

An example will show how these ideas are used in practice. Three solvents were all known to dissolve a particular family of complex organic chemicals. A new compound from that family was developed, and to process it into its final form, it was important to find whichever solvent or mixture of solvents could dissolve the greatest amount of the new compound. The experiment shown in Fig. 17.5 was run.

	A	B	C	(symbolized by)
Run 1	100	0	0	♠
2	0	100	0	♠
3	0	0	100	♠
4	50	50	0	♣
5	0	50	50	♣
6	50	0	50	♣
7	33.3	33.3	33.3	♥
8	66.6	16.7	16.7	♦
9	16.7	66.6	16.7	♦
10	16.7	16.7	66.6	♦

Figure 17.4 Sample worksheet for an augmented simplex design

	MEK	Toluene	Hexane	Solubility (g/l)
Run 1	100	0	0	121
2	0	100	0	164
3	0	0	100	179
4	50	50	0	140
5	0	50	50	180
6	50	0	50	185
7	33.3	33.3	33.3	199
8	66.6	16.7	16.7	175
9	16.7	66.6	16.7	186
10	16.7	16.7	66.6	201

Figure 17.5 Solvent mixture experiment worksheet; MEK, methyl ethyl ketone

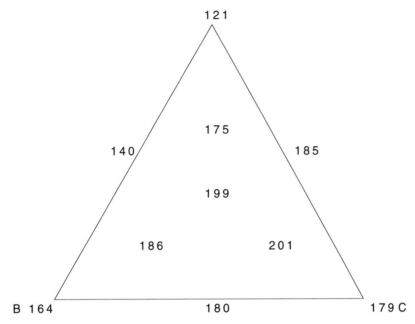

Figure 17.6 Triaxial plot of solubility values for three solvents (A, B, and C). MEK, methyl ethyl ketone

If we plot these numbers on the triaxial map of the process, what is observed is shown in Fig. 17.6.

The first thing that "pops out" of these data is that some combinations of solvents are more effective than any of the individual solvents. Pure MEK, toluene, and hexane dissolve 121, 164, and 179 g/l of the chemical, but several of the combinations dissolve more than 180 g/l. The two highest values lie at the center point (199 g/l) and in the lower right check point (201 g/l). This suggests the strong possibility of an interaction between MEK and hexane, and also possibly between all three solvents.

The use of regression techniques confirms that a model using terms for all three solvents plus the AC and ABC interactions fits the data very well (R-squared >95%) with all terms having very high confidence levels. The model predicted a maximum solubility for a particular solvent blend of 207 g/l, which was so close to subsequent test data as to be within normal scatter for solubility measurements. If the original experiments had included some replicate runs, the estimated range for the optimum solubility would have contained the actual level measured for the dissolving power of the final solvent blend.

For many situations, the use of all the blends in the standard model is impractical. For instance, in experimenting with the cookie recipe in which flour, eggs, and butter are combined, it makes no sense to have experiments with 100% flour, 100% eggs, and 100% butter, or even any of the two-ingredient blends. Also, a cook with any background knowledge of these ingredients would know that the final recipe would contain more flour than either of the other two ingredients, and probably there would be more eggs used than butter.

What this means is that only a restricted portion of all the possible combinations of the three ingredients needs to be investigated. This amounts to selecting a smaller triangular experimental space to work on from within the full possible space. As shown in Fig. 17.7, the new simplex shape is such that there will always be at least 50% flour in the recipes and the amount of eggs will usually be more than the amount of butter.

The use of a restricted experimental space requires some manipulation of the numbers describing the levels, but that type of transformation will not be described here. DOX computer software will usually make such transformations very easy.

The same software also makes it easy to use simplex designs for more than three control factors, and designs containing six or more ingredients are now

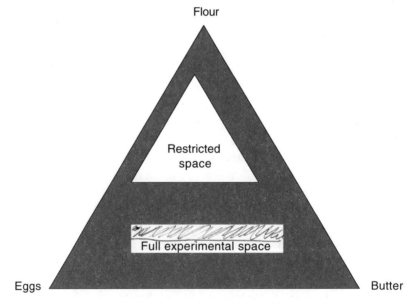

Figure 17.7 Example of a selected experimental area

used regularly. The same principles of simplified mathematical procedures support all mixture designs, but with more than four ingredients it becomes very difficult for most people to visualize the experimental space as conveniently as triaxial graphs do for three-factor designs.

There are times when the same process might be explored with either simplex or factorial designs. The choice of which to use depends on the experimenter's view of the process and perhaps his or her personal preferences. It is also possible to use combination designs in which both simplex and factorial elements are present, but that definitely falls into the area of advanced DOX practice.

The principal items of interest in this chapter are:

- Processes in which the controlling factors are interrelated and must add up to a constant total can be explored using alternate designs that have some mathematical advantages; and
- the simpler designs (three and four factors) can be set up and analyzed without the help of a computer, but in general it is much easier to apply appropriate software to them, and in the case of higher level mixture designs, it is almost mandatory.

18 Latin Squares and Their Derivatives

In the early days of designed experiments, the principal area of investigation was agriculture. The questions of what level of irrigation, which types and levels of fertilizer, which soil variations, which type and particular species of plant, etc., will provide the best yield from a plot of land or for a certain crop are of key importance to the farmer.

In agriculture, the sources of scatter are always a concern; whether there is a fertility gradient or moisture variation across a test field that will lead to erroneous conclusions about a comparison of seed types or fertilizers was a key question for agricultural researchers. (In modern times, it is possible to do more detailed analyses of soil and soil conditions across a field, but before the early 20th century that technology was unavailable.)

One approach to minimizing the effect of such possible sources of scatter on the final data was to break up a test field into alternating plots, organized into columns and rows, so that every tested control factor could be equally exposed to all variations in growing conditions across the field. For example, if four kinds of fertilizer were to be compared, the field would be subdivided into four columns and four rows so that sixteen individual plots were laid out. The different kinds of fertilizer would then be used in the plots so that each fertilizer appeared once in each row and once in each column, as demonstrated in the pattern in Fig. 18.1 (the four fertilizers are designated as A, B, C, and D). This type of design is called a Latin square.

Now, when the average or total yield of plots using one type of fertilizer is compared with the yields from the other fertilizers, the experimenter can feel confident that the comparison is validly contrasting the effects of the fertilizers, with all secondary effects of differences across or down the field being canceled out. This can also be considered as a form of blocking in which the possible variations throughout the test field are blocked out from affecting the data analysis by use of the columns and rows.

The pattern in Fig. 18.1 is not the only possible way to set up a Latin square for four levels of a control factor. Three other setups exist that are perfectly valid, but another 572 possible patterns of the four fertilizers in 16

	Col 1	Col 2	Col 3	Col 4
Row 1	A	B	C	D
Row 2	B	C	D	A
Row 3	C	D	A	B
Row 4	D	A	B	C

Figure 18.1 A 4 × 4 Latin square

	Col 1	Col 2	Col 3	Col 4
Row 1	A1a	B2b	C3c	D4d
Row 2	B3d	A4c	D1b	C2a
Row 3	C4b	D3a	A2d	B1c
Row 4	D2c	C1d	B4a	A3b

Figure 18.2 A 16-run hyper-Graeco-Latin square

plots can be drawn up which do not properly balance out all likely changes in effects caused by variations across the test field.

Mathematicians and experimenters worked on expanding the use of this concept and soon developed Graeco-Latin squares, in which a third noise factor was able to be blocked by inclusion in the pattern. Finally, hyper-Graeco-Latin squares (HGLS) were drawn up, which had a fourth noise factor added in. A four-level example is shown in Fig. 18.2, in which the noise factors are represented by columns, rows, Arabic numerals, and lower case letters, while the control factor being investigated is still a capital letter.

The important properties of these designs are that they are orthogonal, which means they are fully balanced in several senses. Each symbol of any variable (such as the capital letters, the numerals, the lower case letters, or Greek or Chinese characters if used) appears once in each column, once in each row, and once in combination with every other symbol for every other variable. This means that comparisons between the average effects for the main factor levels remain valid estimates of their true effects.

It turns out that, for any size HGLS, the number of factors that can be fitted into each subsection is one fewer than the size (the number of subsections per side) of the square. Thus a 5 × 5 square could have the main experimental factor and three noise factors combined into it, along with the noise factors represented by the columns and rows. The number of possible invalid combinations of all the factors in one of these designs is even higher than for simple Latin squares and, in fact, rises rapidly from thousands to billions for a 7 × 7 square. Fortunately, tables of some of the comparatively few valid designs are available.

A typical problem that might be tackled using these designs could be evaluation of mileage performance resulting from different tire tread designs; the noise factors could be the models of cars used, the individual test drivers, different kinds of driving, etc. There are many situations in which factors known to have an effect on the output of some process are not themselves of interest, such as test machine variations or operator shifts, and the forms of Latin squares are frequently employed to block out the unwanted effects.

However, use of the designs has not been limited to the analysis of just one main factor effect while blocking out multiple noise factors, even though that has always been the formal statistical justification for them. Experimenters soon decided that the contrast of average effects of each factor in any one of the square patterns could be highly useful.

The idea of screening N factors at $N - 1$ levels in each can seem very attractive. The potential efficiency of the designs becomes very high. Table 18.1 illustrates this for the three main HGLS designs.

As always, if it seems to be too good to be true, it probably is. To gain this tremendous efficiency, something has to be sacrificed, which in this case is

Table 18.1 Comparison of HGLS Designs with Full Factorials

Square size	Factors	Levels	Runs	Full factorial	% Used
3 × 3	4	3	9	81	11.1
4 × 4	5	4	16	1,024	1.6
5 × 5	6	5	25	15,625	0.16

essentially all information about interactions between factors. The details of rigorous statistical analysis of data from HGLS designs are not simple, and there are some disputes among statisticians about what techniques are most appropriate.

The most important point for the user is that the designs can be considered reasonable tools for screening multiple factors and factor levels in a process that is still being broadly explored. The major risk involved is that if, in some situation, a major interaction does exist between two factors, let's say A and D, but only one of them (A) has a large single-factor effect, the screening data would indicate that the other (D) could be eliminated from the process or be used at its most convenient level. From that point on, a large, positive AD interaction effect could be lost to the experimenter, or a large negative AD effect might be accidentally triggered, with no clear indications to the investigator of what has occurred.

Depending on what is known of the process, this risk may be unacceptable or the experimenter may feel confident that likely interactions of any importance can be found in later stages of the work. (Also, in the too often encountered situation in which time is exceedingly short, the gain from quickly learning at least some important basics of the process may be so critical to rapid progress that the risk becomes tolerable.)

Randomization is just as important in using HGLS designs as in any other form of DOX, and various schemes are employed for that purpose. One way is to take the formal design from the appropriate table and associate a factor type with each main term (capital letters for factor 1, columns for factor 2, etc.) and factor levels with the series of each main term (A = level 1, B = level 2, etc.). The large design square then is drawn up, showing each combination of conditions for the subunit squares. Finally, the columns are shuffled randomly in order, and then the rows are shuffled, too (or it can be rows first and then columns).

This method is illustrated in Fig. 18.3 for the HGLS example shown in standard order in Fig. 18.2. The pattern of subunit squares in the randomized large square is then used for the actual experimental run order, going left to right or top to bottom.

An alternate method is to simply assign the levels randomly within each factor type so that if capital letters happened to represent the factor of catalyst concentration, the A-B-C-D sequence would not be 1%-2%-3%-4% but, instead, is randomized to 4%-1%-3%-2%, 2%-4%-3%-1%, or any other random sequence. If this is done for each factor, the end effect is as random as shuffling columns and rows. (Some experts recommend *both* randomizing

	Col 1	Col 2	Col 3	Col 4
Row 1	B1c	A2d	C4b	D3a
Row 2	D4d	C3c	A1a	B2b
Row 3	A3b	B4a	D2c	C1d
Row 4	C2a	D1b	B3d	A4c

Figure 18.3 Randomized 16-run HGLS design

levels and shuffling columns and rows, but this may be overkill; how many times must the deck be shuffled before the cards are dealt?)

The only drawback to the randomization of levels is that for factors that do fall in some sort of numerical sequence, such as the example in percentage of catalyst used above, for analysis it becomes necessary to reorder the levels to their normal sequence.

Once the experiment has been performed and the data are available, the next step is to use the simplest form of analysis possible, which is to calculate contrasting averages for the levels of each separate factor. In the 4 × 4 example used so far, that would mean:

averages for each column of column, 1–4;
averages for each row of row, 1–4;
averages for each capital letter factor, A–D;
averages for each numerical factor, 1–4; and
averages for each lower case letter factor, a–d.

Because this is a screening design, the object is to find major differences. This can be done mathematically, but simple graphing is often effective in demonstrating degrees of differences in effects of the various factors. Sometimes even a cursory look at the various sets of averages will reveal significant trends.

In a real situation, the 4 × 4 example was applied to a chemical synthesis that had the following experimental control factors:

columns = time used for reaction (1, 2, 3, and 4 h);
rows = temperature used for reaction (125, 150, 175, and 200 °C);
capital letters = source of main reagent (coded A, B, C, and D);
Arabic numbers = brand of catalyst (coded 1, 2, 3, and 4);
lower case letters = catalyst concentration (1, 1.5, 2, and 2.5%).

The design (unrandomized) is shown in Fig. 18.4.

In this case, the 16 combinations of experimental conditions were all entered on slips of paper that were drawn at random for the actual series of runs. The resulting data for percent yield of theoretical maximum product are shown in Fig. 18.5.

	1 hr	2 hr	3 hr	4 hr
125°	A-1 1%	B-2 1.5%	C-3 2%	D-4 2.5%
150°	B-3 2.5%	A-4 2%	D-1 1.5%	C-2 1%
175°	C-4 1.5%	D-3 1%	A-2 2.5%	B-1 2%
200°	D-2 2%	C-1 2.5%	B-4 1%	A-3 1.5%

Figure 18.4 Example of 16-run HGLS design applied to a synthesis process

	1 hr	2 hr	3 hr	4 hr
125°	54.2	60.6	66.3	75.1
150°	63.4	70.1	77.3	73.6
175°	70.5	79.7	86.8	75.1
200°	86.7	67.2	58.6	64.9

Figure 18.5 Yield results superimposed on the pattern of Fig. 18.4

Table 18.2 Average Responses Across Four Levels of Factor Settings

Time	1 h	2 h	3 h	4 h
	68.7	69.4	72.3	72.2
Temperature	125 °C	150 °C	175 °C	200 °C
	64.1	71.1	78.0	69.4
Reagent	A	B	C	D
	69.0	64.4	69.4	79.7
Catalyst	1	2	3	4
	68.5	76.9	68.6	68.6
Catalyst concentration	1%	1.5%	2%	2.5%
	66.5	68.3	74.6	73.1

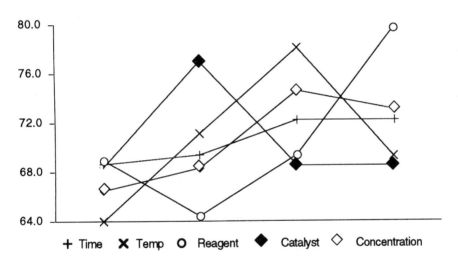

Figure 18.6 Line graph showing effects of five factors on yield

The average yields calculated for the five factors then appear as shown in Table 18.2.

Trends can be seen in simple graph form, as displayed in Fig. 18.6.

With the use of either the table of numbers (Table 18.2) or the graph (Fig. 18.6), a first-glance digestion of these data leads to some observations:

- The effect of time on yield appears to be positive but may level off by 3 h;
- temperature has a more dramatic effect, but yield appears to peak somewhere before 200 °C;

- reagents from sources A and C seem to be equivalent and source B provides lower results, but source D is clearly the best;
- catalysts 1, 3, and 4 are more or less identical, but catalyst 2 stands out with higher yields; and
- increasing catalyst concentration raises yield slowly, but it may level or peak out around 2%.

Either way, it looks as if the process will probably run best at 3-4 h of reaction time, 175 °C, using material from source D with catalyst 2 at a 2% concentration. That particular set of conditions does not appear in the square as drawn up, but when three confirming runs were made using those conditions, their results were narrowly distributed and averaged >90% yield.

Subsequent exploration of the process demonstrated that optimizing temperature, source material, and catalyst alone provided high yield even at significantly lower reaction times and catalyst concentrations, resulting in part from interactions between those three main factors. Fine tuning of the process through response surface methods led to a more efficient combination of conditions (faster and less costly), with a final yield level approaching 95%. However, the initial HGLS work had given needed information on the process in a very timely manner, even though it did not provide the kind of detailed model that a series of factorial designs can generate.

Sometimes the numbers of factors and levels may not match up perfectly into the "square" pattern. For instance, in the case above, there could have been only been two suppliers of reagent or two types of catalyst available, even though all the other factors needed to be checked at four levels. Such a situation does not necessarily make use of HGLS designs impossible.

If there were only two suppliers of reagent, then the experimenter would set A = B and C = D in the design, and a full balance of the design would be maintained. (Of course, in the analysis, only two averages would be compared for the reagent, not four.) Likewise, if only two forms of catalyst were to be tested, they could be set as 1 = 4 and 2 = 3, and the contrasts would still all remain valid.

There are also cases in which matching the available factors or levels to the pattern is not so convenient, such as if there were three suppliers of reagent in the example given. It is still possible to force fit the design with unbalanced assignments within the pattern, but this can make both informal and formal analyses of the data more difficult and less confident.

Not having as many factors to evaluate as the design will accept, such as five factors for a 5 × 5 square instead of six, is not only no problem, it is actually an advantage. The data can still be analyzed as usual, including the contrast of the "empty" factor (also called a dummy variable) assignment; the level of variation in those averages becomes a very useful estimate of the scatter within the entire experiment.

If in the chemical synthesis example there had actually only been one reaction time used, so that the spread of averages for the columns was not directly related to any controlled variation of the process, then that spread (from 78.7% to 82.3%, about a 3.6% difference) could be considered normal. The smallest other difference observed was about 7%, which is enough larger than 3.6% to warrant a high level of confidence that real effects on the process are being detected for all other factors. If there is no dummy variable in the design to permit estimation of normal scatter and some of the differences between averages are not large, then the experimenter may have to rely on prior experience with the process or related processes in deciding what degree of contrast in the averages justifies acceptance of the particular factor as being significant.

Although many different sizes of HGLS designs have been drawn up over the years, the 3 × 3 (Taguchi L9), 4 × 4 (seen above), and 5 × 5 (Fig. 18.7) are by far the most used. Because it is extremely seldom that anyone would need more than the 5 × 5, which allows up to six variables to be explored at five levels each, none of the other patterns will be presented here.

In the 5 × 5 array, there are columns, rows, capital letters, Arabic numbers, and lower case letters, as used before. In addition, Greek letters are used as well for the additional possible factor for this HGLS.

	Col 1	Col 2	Col 3	Col 4	Col 5
Row 1	A1aα	B2bβ	C3cχ	D4dδ	E5eε
Row 2	B3dε	C4eα	D5aβ	E1bχ	A2cδ
Row 3	C5bδ	D1cε	E2dα	A3eβ	B4aχ
Row 4	D2eχ	E3aδ	A4bε	B5cα	C1dβ
Row 5	E4cβ	A5dχ	B1eδ	C2aε	D3bα

Figure 18.7   A 25-run HGLS design

These designs are often misused or overused but are very worth knowing, because their proper applications can be more than convenient when approaching a very new process with multiple possible factors and levels to be considered quickly (particularly when they are to be used, as originally intended, to screen out several noise factors when evaluating a single control factor). The experimenter must remain fully conscious of the risks inherent in utilizing them, but, under the right circumstances, they can be another very valuable tool in DOX methodology.

Summary points are:

- Latin square designs were originally developed to remove the effects of extraneous noise factors from experiments on a primary independent variable affecting a process;
- designs of this type can be used to sort out numerous main factors as well, but then there can be a major risk associated with losing all information on interactions;
- the most compact designs are hyper-Graeco-Latin squares (HGLS), which allow evaluation of up to $N + 1$ factors in an $N \times N$ array; and
- the designs are easiest to use when all factors can be used at N separate levels.

19   Basic Analysis of Variance

Analysis of variance (ANOVA) is a major tool in many statistical procedures and certainly in data analysis from designed experiments. At first glance this sounds inappropriate, because in design of experiments (DOX) the comparisons are usually between the averages of two or more groups of results, not between the amount of variation in the groups (which statisticians prefer to measure by the variance, even though most laymen use the standard deviation or SD). At this point, it is necessary to go into some mathematics.

The variance for any set of n numbers from a sample is the preferred way of examining the spread (formally called the dispersion) of the numbers. It is defined as the sum of their squared differences from their average, divided by $n - 1$. The equation below shows this in formal mathematical notation but is really not as complex as it looks.

$$\text{Variance } (X_1 .. X_n) = \frac{\Sigma(X_i - \overline{X})^2}{n - 1}$$

The symbol Σ just means to add up all the possible combinations of the particular form shown in the parentheses. In this case, $(X_i - \overline{X})^2$, where X_i stands for each separate number in the set of numbers being examined and $\overline{X}$ is the average of the numbers in the set, so that each $(X_i - \overline{X})$ measures the difference between one of the numbers in the set and the overall average of the set. Each difference is squared before they are added up, which eliminates any effect of whether the difference was a plus or minus amount.

Let's apply this to a small set of numbers, for instance the six numbers 6, 4, 7, 5, 3, and 5. These add up to a total of 30, which divided by 6 (the number of different numbers in the set) gives an average of 5 for the group. The way the variance is calculated would lead to

$$\text{Variance} = \frac{(6 - 5)\,2 + (4 - 5)\,2 + (7 - 5)\,2 + (5 - 5)\,2 + (3 - 5)\,2 + (5 - 5)\,2}{(6 - 1)}$$

which then becomes

$$\text{Variance} = \frac{(1)\,2 + (-1)\,2 + (2)\,2 + (0)\,2 + (-2)\,2 + (0)\,2}{5}$$

and then

$$\text{Variance} = \frac{1 + 1 + 4 + 0 + 4 + 0}{5} = \frac{10}{5} = 2$$

The variance for that particular set of numbers is 2; it is worth noting that the quantity 10 is also called the "Sum of Squares," and Sums of Squares are used in many other statistical techniques. The SD for these numbers would then be 1.414, which is the square root of 2, because the SD is the square root of the variance. This means that if the six numbers in the set did belong to a normal distribution with an average of 5, then 95.5% of the numbers in the distribution should fall between 2.2 and 7.8, which is the same as $5 \pm (2 \times 1.4)$. (This all goes back to the material in Chapter 4.)

With the basics about variance established, we can now move on to consider what ANOVA is and how it is used. To fully understand the principles of ANOVA, it is necessary to go further into the background of Statistics, starting with a problem faced by mathematicians in the early 20th century. The question was then how to validly compare two groups of data to determine whether they might both belong to the same general population.

One way to look at this is to see whether their averages are significantly different. A specialized test called Student's t-test was invented by a man named William Gossett to compare averages (or means) of sample groups to demonstrate whether they are different enough to be considered as coming from separate populations. Because in Statistics populations are described not only by their averages but also by how spread out they are (their variance), sample groups might differ from each other either by having either different averages or different variances (or both, of course). For instance, among the three sample groups shown in Fig. 19.1, two have the same averages but different variances, whereas two have the same variance but different averages. (Group A is the same one used in the example of how to calculate variance earlier in this chapter.)

Once Mr. Gossett had developed the t-test, someone else (Ronald Fisher) went on to examine the distribution of variances among normal populations and ultimately drew up a series of distributions applying to variances. These are called the F-distributions.

The way F-distributions are used to compare two groups is simple. First, the ratio of the larger variance to the smaller is calculated. If the two sample groups were from the same population, the variance of each is an independent estimate of the variance of that population, and so they should be relatively close to each other. This means their ratio would approach 1. Because random variation among the individuals in each subgroup will naturally cause each to

	Group A	Group B	Group C
	6	9	7
	4	7	1
	7	7	3
	5	6	5
	3	8	3
	5	5	11
Average	5	7	5
Variance	2	2	12.8

Figure 19.1 Three sample groups

be an imperfect estimate of the population variance, some scatter in the ratio is to be expected.

Fisher's table provides a list of ratios that are the largest ones possible for two variances to have (at a stated confidence level) and still come from the same parent population. However, as with the t-table, the amount of information being used, as degrees of freedom (df), is a part of the determination. Therefore, the df for each subgroup ($n - 1$) must be applied in the use of the F-distribution.

In the example in Fig. 19.1, both Group A and Group C have six elements and so have 5 df each. A statistician would go to an F-table and locate the ratios applying to comparisons with 5 df for the numerator and 5 df for the denominator. They are:

0.05 significance:5.05
0.01 significance: 11.0

Because the ratio of the Group C variance to the Group A variance is 12.8/2 = 6.4, the probability of being wrong if we assume that the two groups come from populations with different variances is less than 5% (which is often referred to as a 95% confidence level that the variances are indeed different). However, because the ratio is less than 11, the probability of the groups coming from the same population is still greater than 1%.

The ratio of 11 may seem high, but two facts must be kept in mind; first, variances of subgroups cover a wider relative range than do their averages, and second, these are not large subgroups, thus the need for a large difference between the variances before the 0.01 significance level is reached. If each subgroup had 15 df, the required ratio for 99% confidence in a real difference between them would drop to the 3.52 level.

If F-tables apply to group variances and the goal of the analysis is to determine whether differences exist between group averages, some special way of applying the F-table to the data must be used. Again, using Group C from Fig. 19.1, the appropriate calculations provide Fig. 19.2

In Fig. 19.2, the sum of the simple deviations from the average is, as it has to be, zero, but the Sum of Squares is a positive number, which in itself gives a measure of how widely distributed the individual numbers are around their mean.

If two sample groups are to be compared to see whether their means are different using ANOVA, the first step is to consider them as one combined group and determine their overall average and Sum of Squares. Using Groups A and B from Fig. 19.1, this results in the calculations in Fig. 19.3.

Thus the Total Sum of Squares from the grand average (SST) is 32 and measures all the variability in the full collection of individual numbers. In this case, each Sum of Squares for the subgroups (A and B) happens to be equal (16), but even if they were not, they would have to add up together to the SST.

However, each subgroup can be examined separately for deviations within itself from the subgroup average. If this is done for A and B, the calculations in Fig. 19.4 result.

The Sums of Squares in Fig. 19.4 represent the amount of variability within each subgroup. Adding them together results in the Sum of Squares Within (SSW), which is a measure of the level of variability in the overall group which is attributable to dispersion within the subgroups. In this case, each subgroup total is the same (10) and the SSW is 20.

Because the SST measures total variability (32) and SSW measures within-subgroup variability (20), their difference (32 − 20 = 12) must be caused by some other sources of variation. One major other source is called

Group C	$X_i - \overline{X}$	Squares
7	2	4
1	−4	16
3	−2	4
5	0	0
3	−2	4
11	6	36
Total 30	0	64 (Sum of squares)

Average ($\overline{X}$) 5

Variance = Sum of Squares/$(n - 1)$ = 64/5 = 12.8

Figure 19.2 Example of variance calculation

Group A	$X_i - \overline{X}$	Squares
6	0	0
4	−2	4
7	1	1
5	−1	1
3	−3	9
5	−1	1
Group B		
9	3	9
7	1	1
7	1	1
6	0	0
8	2	4
5	−1	1
Total 72	0	32

Figure 19.3 Calculation of SST

Group A	$X_i - \overline{X}$	Squares	Group B	$X_i - \overline{X}$	Squares
6	1	1	9	2	4
4	−1	1	7	0	0
7	2	4	7	0	0
5	0	0	6	−1	1
3	−2	4	8	1	1
5	0	0	5	−2	4
Totals 30	0	10	42	0	10

Figure 19.4 Calculation of group SSW total

the between-subgroup variability, and its term is Sum of Squares Between (SSB). The SSW and SSB always add up to the SST when only a single set of comparisons are being made between subgroups. (This is referred to as a one-way ANOVA.)

If two groups were so similar as to really be identical, their averages would be equal to each other and therefore also equal to the grand average, and the SSW would become equal to the SST. All the combined group variation would be attributable to within-subgroup variation and none to the between-subgroup variation, and SSB would be zero.

In the case in Fig. 19.4, SSB is not zero; in fact, it appears to be of significant size compared with the SSW. This indicates that a noticeable degree of the overall variation in the combined group is caused by differences between subgroups. The next question is whether the subgroup contribution to overall variation is large enough to demonstrate that the subgroups are not likely to be from the same parent population. To determine this, each of the Sums of Squares is converted into an average, called the Mean Sum of Squares (MSS).

The SSW and SSB become the MSW and MSB (Mean Squares Within and Mean Squares Between) when each is divided by the df available in the calculation of each. For the SSW there were $n-1$ df in each subgroup, which in the example was $(6-1=5)$, and there were two subgroups, so the calculations become

$$MSW = SSW/\text{total subgroup df} = 20/10 = 2.$$

For SSB there were only the two subgroups involved, so $n-1$ becomes 1 and the equation is

$$MSB = SSB/(n-1) = 12/1 = 12.$$

Notice that the df for MSW and MSB add up to 11; which is the same df used in calculating the SST. These totals must always match up.

Now two numbers are available for comparison, the MSW and MSB, each of which is composed of Sums of Squares divided by df. However, the basic equation for variance shown earlier appears to be in essentially the same format, and in fact the MSW and MSB are two different ways of estimating the variance of the hypothetical parent population of the groups. Also, if they are from the same population, then their ratio will fall within the allowable range in the appropriate F-table.

Therefore, the ratio MSB/MSW is calculated, and in this example it is seen to be

$$MSB/MSW = 12/2 = 6.$$

Examining an F-table for 1/10 df yields terms of 4.96 (0.05 level) and 10.04 (0.01 level). This justifies a confidence level of 95+% that the two subgroups do not really come from the same parent population.

What has happened here is that the separate contributions to overall combined variation made by within- and between-subgroup variation were different enough to not be explainable by random error. Therefore, the between-subgroup variation had to be caused by real subgroup differences.

This succession of steps has taken the examination of the data to the point where use of the F-table, originally meant to compare variances of different groups, can be applied to MSS to reach conclusions about a comparison of subgroup averages.

The particular example is extremely basic; in fact, for two subgroups the comparison would be done just as validly and perhaps more easily by using the t-test. However, that test only applies to comparisons of two groups, whereas ANOVA can be applied to multiple groups and with regard to more than one type of comparison at the same time. The more advanced uses of ANOVA will be dealt with in Chapter 20.

Some cautions have to be registered concerning use of ANOVA in designed experiments. One underlying assumption of ANOVA is a uniform level of variability throughout the explored experimental volume, i.e., the standard deviation of the response is the same for all experimental conditions. This is certainly not true all the time and can affect interpretation of the data.

Also, depending on the signal-to-noise ratio for a given response and the level of replication used in an experiment, it is possible using ANOVA to rate a real factor in some process as statistically insignificant or to locate a statistically significant factor with a contribution too small to be of real concern. This is why the proper sizing of experiments is so important and also why simplistic methods such as the scree plot can be both convenient and useful.

Nonetheless, ANOVA tables are built into all DOX software and are still a key tool. Experimenters can employ the software without reference to the tables, but some will wish to learn more about how the tables are typically drawn up and what the various terms used signify.

20 Full Use of Analysis of Variance

Analysis of Variance (ANOVA) techniques are used in general to examine a collection of subgroups to determine whether they all belong to one homogeneous population. The subgroups are often clearly different from each other, made up of completely separate sets of individual numbers, such as groups of plastic test batches that have all been made with different additives to determine whether the substances have significant effects on flex life or tear strength.

In a typical designed experiment, e.g., an eight-run pattern on three factors, the subgroups that will be analyzed using ANOVA are all made up of the same individuals, which are the eight responses from the different experimental runs. The subgroups all consist of different ways of splitting up the eight responses into two groups of four. Recall that in an eight-run pattern the actual treatments are as shown in Fig. 20.1. (Results in Fig. 20.1 are to be used in an example of how analysis is done.)

The first subgroup from fig. 20.1 would relate to the effect of Factor A, which is seen in the contrast of the results of runs 1–4 (all minus settings) versus the results of runs 5–8 (all plus settings). The second subgroup would, in turn, contrast results from runs 1, 2, 5, and 6 (all minus settings for Factor B) with those of 3, 4, 7, and 8 (all plus settings). Five other contrasts can be

	Factor A	Factor B	Factor C	Results
Run 1	–	–	–	3.2
2	–	–	+	3.5
3	–	+	–	1.6
4	–	+	+	1.3
5	+	–	–	0.7
6	+	–	+	0.8
7	+	+	–	0.6
8	+	+	+	0.3

Figure 20.1 Classic 3×2 full-factorial design with one response

examined, relating to Factor C and to the AB, AC, BC, and ABC interactions, using exactly the same pattern of pluses and minuses derived for their effects.

The normal analysis of effects, using addition and/or subtraction, etc., as shown in Chapter 8, would result in the numbers shown in Fig. 20.2, but ANOVA is a different procedure. The first step would be to calculate the total Sum of Squares (SST), which was demonstrated in Chapter 19. In this case, the grand average is 1.5, so the SST is calculated as shown in Fig. 20.3.

The SST is a measure of all the variation in the data set. If the amount of variation within any subgroup is close to the SST, the subgroup is not significantly different from the overall date set.

In examining any of the individual factors, a set of steps would be taken. Factors B and C will be used to demonstrate.

First, the data would be split up into groups according to the sign given to them in connection with the individual factor. In the case of B, the − numbers were 3.2, 3.5, 0.7, and 0.8 (average = 2.05), whereas the + numbers were 1.6, 1.3, 0.6, and 0.3 (average = 0.95). The actual calculations are shown in Fig. 20.4.

From Fig. 20.4, the sum of 6.80 and 1.10 gives a total of 7.90 for the subgroup called the Sum of Squares Within (SSW), which is much less than the full SST of 10.32, so the amount of variation in the Factor B subgroup is less than that of the whole data set. The difference, in the amount of 2.42, is

Factor	A	B	C	AB	AC	BC	ABC
Effect	−1.8	−1.1	−0.05	0.8	−0.05	−0.25	0.05

Figure 20.2 Summary analysis of effects in Fig. 20.1

Results	$X - \overline{X}$	Squares
3.2	1.7	2.89
3.5	2	4
1.6	0.1	0.01
1.3	−0.2	0.04
0.7	−0.8	0.64
0.8	−0.7	0.49
0.6	−0.9	0.81
0.3	−1.2	1.44
Average = 1.5	Total = 0	Total = 10.32 (SST)

Figure 20.3 Calculation of SST for all results from Fig. 20.1

– Group	$X - \bar{X}$	Squares	+ Group	$X - \bar{X}$	Squares
3.2	1.15	1.32	1.6	0.65	0.425
3.5	1.45	2.10	1.3	0.35	0.125
0.7	−1.35	1.82	0.6	−0.35	0.125
0.8	−1.25	1.56	0.3	−0.65	0.425
	Partial SS = 6.80			Partial SS = 1.10	

Figure 20.4 Calculation of partial SS for Factor B

– Group	$X - \bar{X}$	Squares	+ Group	$X - \bar{X}$	Squares
3.2	1.675	2.81	0.7	−0.825	0.681
3.5	2.025	4.10	0.8	−0.675	0.456
1.6	−0.075	0.005	0.6	−0.925	0.856
1.3	−0.175	0.030	0.3	−1.175	1.381
	Partial SS = 6.94			Partial SS = 3.37	

Figure 20.5 Calculation of partial SS for Factor C

a measure of how much the Factor B subgroup varies from the grand average and is called the Sum of Squares (SS) for B.

When the same procedure is used for Factor C, the corresponding analysis looks like Fig. 20.5.

When carried out to three places, the subgroup SSW total is 10.315, which is very close to the SST of 10.32, demonstrating that Factor C has associated with it the same overall amount of variation as was found for the entire data set. Therefore, Factor C has no real effect on the process result and the SS of only 0.005 for C shows the lack of significance of this factor.

If this procedure is followed for all three factors and every interaction, the seven individual SS values calculated will add up to the same total as the SST. However, each calculation of an individual SS uses one degree of freedom (df), so all seven df available in an eight-run experiment will be used up. This means there is no information remaining to be used in calculating any sort of error term for the equation describing the relationship of the response to the control factors and their interactions. In that case, the use of the full interaction model, which for three factors requires seven terms, has resulted in exhausting all the df available.

The ANOVA table for the full model in this example would usually be displayed in the sort of form given in Fig. 20.6.

	Sum of Squares	df	Mean square	F#	Prob. > F
Model	10.320	7	1.474		
Residual	0	0			
Total	10.320	7			

Root MSE		R-Squared	1.00
Dependent mean	1.50		

Figure 20.6 ANOVA table for experiment in Fig. 20.1 (MSE, Mean Square Error. Dependent mean is the simple average of all the data.)

Because the full model describes all the data perfectly, R-squared is equal to 1, but there is no residual SS, and no F-ratio or error term (MSE) can be calculated. (Note: this does *not* mean the model is perfect!)

Replicates of some experimental runs or even runs made at some intermediate factor settings (such as a center point) can be made, which then supply extra information useful to calculating various kinds of error terms.

Alternately, something less than the full interaction model might be used, which would free up some information for error analysis. The justification of dropping terms from the model is usually made by considering which factors or interactions have comparatively small SS contributions. In the case of the example, A, B, and the AB interaction have the clearly larger SS contributions, and a model using those terms only can be used. How well the reduced model still describes the process is shown by the calculated R-squared term (see Chapter 16). When R-squared remains high, the decision to go to the reduced model is supported.

The terms that are freed up by use of a reduced model can then be used to calculate a combined SS, which is known as the residual SS (RSS). The ratio of any other MSS being used in the model to the mean RSS is another ratio of variances and can be examined by using an F-table to calculate the probability of two such variances occurring together by chance. When the ratio is high, the probability becomes low; this indicates that the particular comparison of an MSS to the RSS demonstrates that the MSS is describing a significant term in the equation, which translates to a real effect on the process described by the equation. The ANOVA table for the reduced model then appears as shown in Fig. 20.7.

Note how the df have been reallocated and that R-squared has been reduced only slightly. The F-ratio shows that the model has a very high probability of being a valid description of the process, because the assumption that the mathematical fit achieved by it was only a result of random

	Sum of Squares	df	Mean square	F#	Prob. > F
Model	10.18	3	3.393	96.95	0.0003
Residual	0.140	4	0.035		
Root MSE	0.1871		R-Squared	0.9864	
Dependent mean	1.50				

Figure 20.7 ANOVA table for reduced model from Fig. 20.6

chance has a very low probability of being correct (~0.03%). Therefore, the model

$$Result = 1.5 - 1.8A - 1.1B + 0.8AB$$

is a very good one for the process being examined.

In the reasonably simple example used, the original calculations of factor and interaction effects showed right away that A, B, and the AB interaction were the main contributors. A scree plot of the effects would also quickly demonstrate which three are significantly higher than the other four. Also, there are mathematical short cuts that use the effect levels to calculate the individual SS terms conveniently, without going through the more laborious steps shown above.

The typical use of ANOVA in response surface method analysis includes the generation of more data points than are needed to develop a particular model of the process. This can be done in two ways. The first is simple replicates of one or more sets of experimental conditions, as described in Chapter 15, which makes possible estimation of the normal scatter of results in the process. This is referred to as "pure error."

Other unreplicated sets of experimental conditions are also available as information (df) for use in calculations. For instance, the number of coefficients that have to be calculated in a full quadratic model of a three-factor process is nine. They are shown in the equation below as all the C_i terms.

Quadratic model
$$Y = Intercept + C_1X_1 + C_2X_2 + C_3X_3 + C_{12}X_1X_2 + C_{13}X_1X_3 + C_{23}X_2X_3 + C_{11}X_{12} + C_{22}X_{22} + C_{33}X_{32}$$

Thus at least 9 df, from 10 different experimental runs, are necessary for a good analysis and development of this equation.

However, a central composite design for three factors contains 15 experimental runs, even aside from the usual replicates of the center point. This means that five extra df are available.

Calculations can be made based on these extra df to see how well the results calculated by the model fit with the actual recorded data from the experiment. The resulting number is the lack-of-fit term, and if the model is really a good one, this number will be low.

Also, if the model is really good, the terms for lack of fit and pure error may be measuring the same general scatter in the experimental data. The ratio of their mean squares then will not be very high.

However, if the pure error term is small and the lack of fit term is large, the data are showing that the overall differences in the results predicted by the model from the actual data are difficult to explain by normal scatter in the test results. This tells the experimenter that there are other sources of scatter in the process which the derived model does not explain very well.

Figure 20.8 is an example of a full ANOVA for a complicated process.

In the case in Fig. 20.8, the high RSS (2042.5) compared with the model results in a low F-ratio, which implies about one chance in four that this model is inappropriate. This would immediately concern the investigator, as would the low R-squared term.

However, when the RSS is broken up into the terms for lack of fit and pure error, it is immediately obvious that the scatter detected in the replicates is actually quite small. The great bulk of the RSS comes from the term for lack of fit, and the ratio of lack of fit to pure error is a large number. This leads to a very high F-ratio, which implies a very low probability that the overall difference between the real data and the predictions of the model is related to simple experimental error. This strongly reinforces the investigator's drive to go back and develop a better model.

In other cases, there might be a good overall model fit to the data and a high R-squared, but the term for pure error might actually be larger than that for the lack of fit. If their ratio is far enough from 1, the data are telling the experimenter that, even though the model fits the data nicely, there was a possibly atypical amount of simple scatter among the replicates. This may

	Sum of Squares	df	Mean Square	F#	Prob. > F
Model	6750.2	9	750.03	1.836	0.2607
Residual	2042.5	5	408.49		
Lack of Fit	2040.3	3	680.10	627.8	0.0016
Pure Error	2.2	2	1.08		
Root MSE	20.21		R-Squared	0.7677	
Dep. Mean	47.16		Adj. R-Squared	0.3496	

Figure 20.8 ANOVA table for a complex process

cause some re-examination of the process or the testing to see why so much basic experimental error is being detected.

Even after a model has been selected, an ANOVA table has been constructed, and some of the numbers appear to support the model as being appropriate, there are ways to check on possible problems with the data. All statistically based techniques rely on some assumptions about the data being examined, and if any of those assumptions are not met, the validity of the analysis comes into question.

In the case of ANOVA, some of the assumptions are about the randomness of experimental errors, lack of bias in the data, etc. Various techniques are used, usually in the form of simple graphs, to make sure that there are no nonrandom patterns in the data which point to some sort of flaw in how they were generated. For instance, if a plot of residual error versus order of experimentation showed a clear trend upward or downward, it would raise whole new questions about the way the experiments were run.

Every DOX software package will have a number of routines built in for examining how good the models are and how well the data or model predictions match up against various criteria. Proper use of some of these will take more study of the underlying statistical principles than falls within the scope of this book. The experimenter is advised to rely first on the simpler tools available in the software analytical functions and then, if some concern arises about possible problems with the data, to carefully study the manual before progressing to the more advanced tools of the program.

21   Taguchi's Contributions

Dr. Genichi Taguchi has been the major driving force behind the general adoption of design of experiment (DOX) methods in Japan, and in the early 1980s his teachings came into vogue in the United States with Ford Motor Co. as the principal supporting organization. The overall value of his contributions is undeniable, but there has been a certain amount of controversy over his methods, fueled initially by the strong objections of classical statisticians to one of his techniques in particular. Subsequently, other disagreements arose as a few Taguchi disciples not only spurned the objections to that technique but more or less claimed that Taguchi methods and philosophy in some senses superseded much of classical DOX practice altogether.

Defensive reactions against that thinking became at times intense, and some very vigorous debates occurred in the later 1980s. There are still a few people who in general dislike and/or refuse to use any Taguchi methods and, also, some Taguchi adherents who feel that the only way to approach any experimental design is through use of those methods.

Authorized U.S. instruction in these methods is provided by the American Supplier Institute in Detroit, MI, headed by Dr. Taguchi's son. Courses start at two weeks in length and can be extended to four weeks, which is much longer than any other form of nonuniversity instruction in DOX commonly available. This is because the instruction is heavily focused on philosophy and the development of special concepts of process optimization and tolerance design as well as on actual experimental patterns and their interpretation.

The concept of loss function is a cornerstone of the philosophy. This involves two ideas, the first being that less-than-perfect quality in any product always imposes a cost of some sort on society as a whole which the experimenter is ethically obliged to minimize. The second idea is that quality of a product is not equal as long as the product falls within the limits of some specification but actually is better when the product characteristic is narrowly distributed around the specification mean. This is illustrated in the characteristic quadratic loss function diagram originated by Dr. Taguchi, shown in

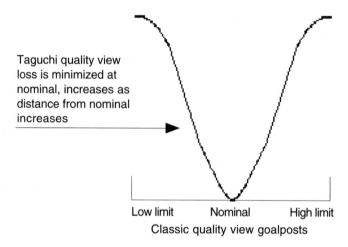

Taguchi quality view loss is minimized at nominal, increases as distance from nominal increases

Low limit   Nominal   High limit

Classic quality view goalposts

Figure 21.1 Comparison of Taguchi loss function and classic quality concept

Fig. 21.1 along with the view of traditional "goal post" quality still used for the most part in U.S. industry.

Thus the experimenter is supposed to carefully examine product performance in the much larger context of overall cost to society, and optimize that performance with these social considerations in mind. Furthermore, process optimization is not to focus exclusively on achieving the average best level of response but to focus also on its narrowest distribution. The latter idea is generally quite compatible with customary Western business methods, but in today's highly competitive atmosphere, few companies are actually willing to subordinate considerations of producibility and immediate customer satisfaction or profit to that of how well the total society will benefit over time from the product.

The special concepts of process optimization revolve around a set of formulas that Dr. Taguchi developed which evaluate both how closely a response approaches its target level and the amount of variability in the response. He called these formulas "signal-to-noise ratios" (a term and concept that are applied in many other areas but with somewhat different meaning), and there are a substantial number of them to be used in various situations. However, the three main ones are the smaller-is-better, larger-is-better, and nominal-is-best formulas. As the names indicate, they are meant to be used in situations in which the ideal response is zero, goes to a maximum, or varies around some set target level.

Actual application of the formulas presents no serious difficulties, but a certain amount of mathematical analysis is involved. Outside of Japan, the majority of statisticians and users of DOX find it more convenient and at least as valid to simply do two separate analyses, one of the average response and the other of the variation of the response.

On some occasions, examination of several process control factors will reveal that a few of them (e.g., A, B, and C) have significant direct effect on the process output average, whereas others (arbitrarily designated D, E, and F) act to broaden or narrow the distribution of the response but have little effect on the average response. In this happy circumstance, the experimenter can then use A, B, and C to tune the process output toward its desired optimum while, at the same time, adjusting D, E, and F so that the process output is reduced to a minimal overall range.

This is an idealized situation and cannot be considered as an expected norm. Quite frequently, either the same factors that affect the process average also affect the spread of the response or the data do not show any dramatic trends in the standard deviation of the response. This is particularly common when a comparatively limited number of replications of runs or product tests are used; because variation in sample standard deviations is generally larger than variation of sample averages, either a substantial contrast between sample standard deviations or a large sample size is required before differentials can be detected with confidence. Despite these concerns, the analysis of process variation is clearly worth attempting, and, when probable factors for controlling that variation are found, the experimenter gains an understanding of a very important facet of the process.

This can be particularly true when a complex process has more than one region of favorable process output. Suppose a product had two different peaks of desired result, for instance, a formula for a plastic that might have a maximum tensile strength of 5,000 pounds per square inch (psi) using one combination of ingredients but 4,800 psi at a very different combination. If the formula that yields 5,000 psi also leads to a broad variation in results (say ±5%) but the formula yielding 4,800 psi has much lower variation (±2%), then the second formula may actually be preferable for use.

A combination of control factor levels that not only brings process results up toward an optimum but, in particular, an optimum that is relatively insensitive to minor changes in factor levels is a notable goal of Taguchi philosophy. Any process that runs to approximately the same favorable level of output regardless of slight drift in running conditions is termed a robust process. Having a robust process is very advantageous for any organization, because it has benefits for everyone—management, sales, and production personnel.

As a basic example, consider a chemical process, the yield of which is temperature dependent and is also affected by the type of catalyst used.

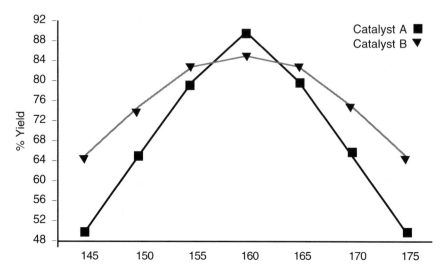

Figure 21.2 Effects of catalysts and temperatures on yield

Suppose a study was done that examined the effects of catalyst and temperature on yield, and the resulting contrast plot looked like Fig. 21.2.

In the case in Fig. 21.2, it can be seen that, although catalyst A does result in the absolute highest yield (90%), the lower peak yield (85%) resulting from catalyst B may be more than made up for by the obviously lower sensitivity of the whole process to reaction temperature. If consistent temperature control were difficult or expensive to arrange, the overall efficiency of the process using catalyst B under actual factory conditions might well be more attractive than the higher but more easily lost peak of the process using catalyst A. The B process is clearly the more robust of the two.

Another of the distinctive characteristics of the Taguchi approach is to consider factors that affect a process as coming from two possible categories. The first group are those factors that are directly operator controllable, which are designated as control factors. Other influences on the process which cannot be controlled at all, to only a limited extent, or through great expense are called "noise factors." Examples might be ambient humidity, slight tool wear, and normal variation between a number of similar pieces of production or test equipment.

One way to improve process robustness is to set the control factor levels so that they minimize the effect of noise factors on the process output. In a sense, this means searching for favorable interactions between the control factors and the noise factors so as to lessen the direct effect of the noise.

In setting up an experimental design, referred to by Taguchi as an orthogonal array, first the array for the control factors is selected and then an overlay of additional conditions based on the noise factors is added. For instance, if four main factors and three noise factors were being investigated, the principal design (the inner array) might be an eight-run half factorial, whereas the noise factor overlay (the outer array) could be a four-run half factorial. Thus each of the eight conditions in the inner array would be run at each of the set of four overlay conditions in the outer array so that a total of 32 actual experimental runs would be drawn up.

This setup is illustrated in Fig. 21.3 with control factors A, B, C, and D and noise factors N1, N2, and N3.

	Inner Array				Outer Array		
Run	A	B	C	D	N1	N2	N3
1	+	+	+	+	+	+	+
2	+	+	−	−	+	+	+
3	+	−	+	−	+	+	+
4	+	−	−	+	+	+	+
5	−	+	+	−	+	+	+
6	−	+	−	+	+	+	+
7	−	−	+	+	+	+	+
8	−	−	−	−	+	+	+
9	+	+	+	+	+	−	−
10	+	+	−	−	+	−	−
11	+	−	+	−	+	−	−
12	+	−	−	+	+	−	−
13	−	+	+	−	+	−	−
14	−	+	−	+	+	−	−
15	−	−	+	+	+	−	−
16	−	−	−	−	+	−	−
17	+	+	+	+	−	+	−
18	+	+	−	−	−	+	−
19	+	−	+	−	−	+	−
20	+	−	−	+	−	+	−
21	−	+	+	−	−	+	−
22	−	+	−	+	−	+	−
23	−	−	+	+	−	+	−
24	−	−	−	−	−	+	−
25	+	+	+	+	−	−	+
26	+	+	−	−	−	−	+
27	+	−	+	−	−	−	+
28	+	−	−	+	−	−	+
29	−	+	+	−	−	−	+
30	−	+	−	+	−	−	+
31	−	−	+	+	−	−	+
32	−	−	−	−	−	−	+

Figure 21.3 Taguchi design with four control factors and three noise factors

Analysis could then proceed initially by looking at the effects of noise using the simple analysis of a four-run half factorial of the three noise factors. (Outer arrays are usually fractional factorials with substantial confounding of two-factor interactions, based on the assumption that noise factors seldom have substantial interactions with each other and it is their main effects that are of primary value and interest.) In that analysis, each subgroup is made up of eight data points relating to the eight run conditions of the inner array.

The inner array can then be examined using the appropriate calculations for that design; in this analysis, each subgroup has four data points relating to the run conditions of the outer array. Often, in the inner array, more consideration is taken of where interactions of the control factors may be located so that they can be specifically examined.

The use of this technique does not normally call for direct evaluation of the interactions between control factors and noise factors. If the investigator wished to do so, he or she would have to look at the overall 32-run design with the total of seven factors (4 control, 3 noise) in it. A full factorial for seven factors requires 128 runs, so the actual experiment would be a 1/4 fractional factorial (in this case, it can also be referred to as a 2^{7-2} factorial). With seven factors involved, there would be twenty-one possible two-factor interactions, plus the seven main effects, for a total of 28 effects to be found and analyzed.

This is possible with 32 total runs, because there are 31 df, 28 of which are needed to calculate all the effects. However, for this to work, both of the two original arrays, inner and outer, would have to be set up so that none of all the possible two-factor interactions is aliased. If the settings of any one column are not exactly correct, then the full pattern of two-factor interactions becomes complicated and some interactions cannot be evaluated by themselves.

If the inner array had been a reduced fractional factorial, below half-factorial size, its combination with the outer array would have inescapably resulted in aliasing of various two-factor interactions. This facet of the inner-outer array technique needs to be understood; when both arrays are fractional factorials, as is often the case, it may easily happen that some two-factor interactions cannot be independently analyzed. If the assumption of nonexistence or insignificance of such interactions is correct, there is no problem, but that assumption is not always a safe one.

The division of factors into control and noise categories is as much a matter of convenience as anything else, which is also true for the use of inner and outer arrays. Some experimenters prefer to consider all factors thought to affect a process and needful of investigation, whether they are directly controllable or environmental, without any particular distinction among them when drawing up a designed experiment. Consideration of which interactions are believed to exist or thought to be especially important can then be taken into account and worked into the design.

In a very real sense, there is seldom a single "right way" to explore a process with enormously higher efficiency than all other ways; any technique or sequence of experiments that allows the investigator to validly contrast factor effects and interactions and gain understanding of the process is perfectly legitimate.

Dr. Taguchi's techniques and philosophy involve a number of different points and have been described in some lengthy books, but a capsule summary of his ideas on basic experimental design practice would be as follows:

- Use saturated fractional factorials (L4, L8, L12, L16) for two-level screening designs as much as possible, and when examining factor effects for significance, use the terms from the obviously smaller factor evaluations as if they were error measurements;
- use three-level screening designs (L9, L18, L27) on occasion, especially when contrasting discrete factors;
- if at all possible, use several replicates of all runs or at least several tests of product from each run to generate test data that not only include the average of the test property but also the standard deviation of each sample group;
- examine the process to see what influences are operator control-lable and which are environmental, and set up a combination of inner array for control factors and outer array for noise factors for the overall experiment;
- use the normal DOX techniques to analyze not only the response (or its average) but also the spread of the response, as indicated by the standard deviation or its logarithm; and
- seek knowledge of the full process so that it can be optimized for the best overall combination of output, consistency of output, and tolerance of variation in factor levels.

22   Advanced Topics in Design of Experiments

The development of the basic types of experimental designs other than full factorials is reasonably easy for most people to understand. They are, after all, various forms of balanced subsets of the full factorials, and many of them can be visualized without too much trouble when three factors are involved. Naturally, the use of four or more factors makes visualization much more difficult, because very few individuals can mentally picture a pattern in more than three dimensions, but understanding how a three-factor central composite or Box-Behnken or even a four-factor simplex design works generally makes people feel comfortable with those types of designs, even when they are applied to larger sets of control factors.

However, more recently some new approaches have been taken in designs for response surface methodology (RSM), which resulted in what are referred to as optimal or algorithmically derived designs. These are not completely unrelated to the older design families but were developed to be even more efficient and effective in exploring an experimental space. Of the several families of optimal designs, the d-optimal is the best known and most often used.

A full d-optimal pattern is larger than the conventional RSM pattern for the same number of control factors, but it is always a reduced version that is used, which is smaller than the RSM array. For instance, one way of developing the full set of d-optimal candidate runs for three factors and a quadratic model adds up to 35 runs, compared to 15 (plus 5 replicates, for a total of 20 runs) in the standard central composite design. However, in that particular method, only 17 actual runs of the total 35 are needed to provide all the information (degrees of freedom) necessary to fit a quadratic model plus allow lack of fit and pure error calculations.

Optimal designs work through an algorithm that starts out by choosing one run from a very basic set of experiments and then adds more from the full optimal set, recalculating the level of added information each added run brings to the design. Once the number of runs is sufficient to allow calculation of necessary terms to describe the process and the value of the next additional run is significantly greater than that of one of the treatments already in the design, the less valuable run is dropped. After enough iterations of adding and dropping the various runs from the large candidate set, the final design is maximally efficient according to the criteria of the algorithm used. [These iterative calculations are mathematically intensive, so optimal designs are used almost exclusively through software for design of experiments (DOX).]

For four factors, the full d-optimal set includes 97 runs, the central composite has 30, (25 plus 5 replicates), and the d-optimal subset commonly used has only 24 runs. This is still not the smallest design applicable to the situation; a variation of the central composite pattern called the Hartley design could be used, which could be performed with 20 runs (17 plus 3 replicates). However, such a minimalist design is marginal in its capacity for error estimates and is also quite vulnerable in its analysis to even a single outlying data point.

As noted in prior instances, there is always a trade-off in selecting a trimmed-down design pattern versus having a high level of confidence in the accuracy and precision of any conclusions reached through digesting the data it generates. When the experimenter has reason to believe that the region of experimental space to be explored is either the most practical one in which the process can be run or is very likely to contain the optimum process settings, the choice of a more detailed design is indicated. That way, much better data and the development of a good working model of the process will result, and only a small set of confirming experimental runs will be needed afterwards, possibly none at all.

Sometimes the investigation of a process will arrive at a stage that calls for an RSM approach while there is still some question as to where the best region lies in a large possible experimental space. The use of a Hartley design or the factorial section only of a central composite (that is, without all the star points and with only one-half the center point replicates) can then serve as a sort of exploratory RSM approach. Depending on the results, the design either can be filled in by running a second block in the same region or it may show that a different area of experimentation should be explored and where that area lies. Often at least some of the runs from the first approach can be used as a base for the next pattern of runs.

In mixture designs there are also a variety of designs from which to choose, although the differences between them are often less obvious than in the case of the different factorial RSM designs. In fact, sometimes even selecting different mixture design options will lead to identical or very similar designs in the end.

The new practitioner of mixture designs is advised to start with the simplest options and then move to more advanced models while comparing them to the first set. When the number of runs is more than adequate for fitting the anticipated model (most typically quadratic for processes of any potential complexity) and the actual experimental treatments make sense to the experimenter, the decision can be made to try the particular design. With time and experience will come better understanding of what the various options in mixture designs entail, and individual preferences for certain designs based on the special technical field may arise.

A major advantage of computer software packages in using mixture designs is their ability to make it easy to draw up designs in a restricted experimental field. In the example in Chapter 17 in which one ingredient (flour) would always be used for at least 50% of the mixture and the other two (butter and eggs) would not be used at the same maximum levels, drawing up the working experimental matrix by hand would be a tiresome exercise. However, most software will allow the user to make very simple entries of what real levels are to be used and then will do all the necessary recalculations in a manner both convenient and invisible to the investigator. This makes use of mixture designs much easier, particularly when a higher number of control factors (4–6) is being examined.

The choices any individual experimenter will make in choosing designs, whether factorial or mixture types, depend most heavily on what is thought to be known already about the process, its major control factors and their appropriate levels, past familiarity with the various available designs, personal preferences in experimental strategies, and the particular software package being used.

It is possible to actually make up combined designs, that is, ones in which some factors make up a mixture together but in which others are independent and would normally be varied in a factorial design. For example, suppose it was a chemical reaction that was of interest, in which a solvent and two reactive chemicals always made up a mixture that was then heated at different times, temperatures, and pressures to yield some product. The levels of solvent and reagents would be treated as a mixture, whereas the time, temperature, and pressure would be varied in the pattern of common factorial designs.

This type of combined design has not been used very often in the past, because of the degree of difficulty involved in drawing them up and analyzing the results, but it is certainly more efficient to use them in special situations such as the example above. Now a few of the more advanced DOX software packages are set up for making combination designs, and their use is increasing.

One of the important aspects of any advanced design is the use of replicate runs to provide information on what level of scatter in experimental results is more or less normal for the process and the test procedure. The majority of the standard patterns use the center point of the design for all replicates, with it being run typically 4–6 times. The spread of the data from these replicates is used to calculate the pure error for the process, which is very helpful in understanding how good the model developed from the data is and how precise the model predictions are.

The choice of the center point for all the replications is based on the assumption that the center of the experimental space is a reasonable place to measure the amount of process variability. (It is also mathematically convenient for the calculation of how the standard error of the determinations changes across the whole experimental volume.) However, process variability cannot be guaranteed to remain relatively constant over the whole experimental region or even to change in just the manner described by some commonly accepted equations dealing with standard error as a variable. In fact, for many processes, the variability is known to change significantly depending on what the settings of the control variables are. (The assumption of nonconstant process variability related to control factors is actually a key facet of the Taguchi approach.)

Therefore, some experimenters choose to replicate runs other than the center point, usually making one replicate each of several points spread out across the pattern of runs. Optimal designs use this tactic instead of replicated center points. The spread of results in each pair of replicated runs is used in a mathematical pooling of all detected variation among the replicates to make a slightly different estimate of pure error. The choice to estimate pure error from multiple replicates of the one point in the center of the experiments or a combination of single replicates of several points distributed throughout the experiments is up to each experimenter, but the need for a total level of replicated runs sufficient to give a good estimate of pure error applies in any case.

Another option open to the experimenter is whether to adjust RSM models to reduce the number of terms applied in them. One school of thought is to begin with the simplest model available, the linear, and see whether the lack-of-fit and R-squared terms are good enough to validate the model and account for the bulk of the variability in the data. If that is the case, the analysis can stop there, but if either or both terms are not satisfactory, then the next model, the quadratic, is examined, and after that the cubic model can be checked if necessary. (It must be noted that the cubic model can only be used properly when the experimental design was built to support it, which is generally not the case for the common RSM designs.)

If no model has a good fit or a reasonable R-squared term (at least 40–50%), then the experimenter can draw the conclusion that either some other unknown factor(s) is affecting the process or there may be a serious problem with the measurement system used to check the results. Alternately, the control factor levels might not be different enough to provide a change in the response big enough to stand out clearly from the background scatter of the process. Possibly, the distribution of the process output is not normal, and a transformation of the data will be necessary before a good regression analysis can be accomplished and a valid model found.

Often, one of the more detailed models such as the full quadratic will provide excellent fit and an acceptable R-squared. If the model is examined in detail, some of the terms may show up as not being too significant. (See the example in Chapter 16.) The model can then be trimmed of clearly extraneous terms to produce a customized equation for the particular response which still has excellent fit and an R-squared that is only slightly reduced from what it was for the more complex model. There is a quantity, called the "adjusted R-squared", that takes into account the number of terms in the model; this number can actually increase as extraneous terms are removed.

Some statisticians believe that models should only be trimmed when none of the entire class of terms of a certain type show any significance; that is, if no two-factor interactions are significant, then all of those terms can be removed from the model. If no squared terms are important, then they can all be eliminated. In a three-factor model, if the AB interaction were large and statistically different from zero, the AC and BC interaction terms would be retained, even if they were very small and their confidence intervals overlapped zero; likewise for the A^2, B^2, etc., terms.

Other models allow elimination of one of a class of terms, such as the AB interaction, but then both the A and B terms have to be retained in the model even if either or both of them do not display significance by themselves. Likewise, if the A^2 term is significant but the A term does not appear to be so, the A term should still be kept in the model. (This is referred to as maintaining proper hierarchy and is important to the functionality of the overall model.)

When using simple factorial experiments, the goal is to find the most important factors and interactions, so eliminating interaction terms to simplify the model is very appropriate. In RSM designs, the goal is to develop the best-fitting model without obviously extraneous terms, so eliminating minor interaction terms is not necessarily important. However, if the data show that a factor does not appear in any significant term in the model, whether linear, interaction, or quadratic, then elimination of that factor from the model will certainly be an improvement.

It is true that retention of the apparently insignificant terms in the model usually has very little effect on the predicted results calculated by its use, so whether to trim the model might be considered a moot point by the layman. Most DOX software will permit the customization of models, but some programs will flash warnings to users if they trim the model in a way the software writer considers inappropriate. Although the individual experimenter can always decide what principles to apply in model choice and fine tuning, it is often just as easy and certainly mathematically more rigorous to allow the program to maintain model hierarchy.

There is a specialized technique that can be used for running designed experiments using a current production process while having minimal impact on the process output. It is called evolutionary optimization (EVOP) and amounts to laying out minimal experimental patterns, typically four- or eight-run, that use factors known to affect the process but with only small changes in level. This is a major contrast to the normal use of bold limits on factor levels, but it largely removes the risk of having the process run into a region of nonfunctionality. Because making scrap is not an attractive proposition to production managers, EVOP is about the only way that real cooperation can be obtained if experiments are to be done on a fully committed production line.

The normal DOX methods involve the making of very few experimental runs that generate extremely good data, that is, that are as accurate and precise as possible. In EVOP the data will be much more scattered, because they come from a relatively uncontrolled factory environment, and because the range of levels used is narrow as well, the signal-to-noise ratio of the anticipated or desired change in the process to its everyday scatter becomes very unfavorable.

However, in EVOP the extremely good data from a very few experimental runs is substituted by scattered data from a large number of production runs instead. When the amount of data used in an investigation becomes large enough, a poor signal-to-noise ratio can be overcome.

In a typical case, there might be a factory injection molding process, the scrap rate of which is known to be affected by the temperature, time, injection rate, and holding pressure. (In this technology, the existence of interactions between the factors would be considered likely.) The process has been tuned enough to run the 36-cavity mold at an objectionable but temporarily bearable scrap rate of 5%, and it is running constantly to work off a backlog of overdue orders. The drive to reduce scrap to 1–2% is high, but all experimental work has to be done during production runs and any dramatic jump in scrap would trigger various kinds of unpleasantness for the experimenters.

In this case, an eight-run half factorial could be drawn up for the four factors, with each being changed minimally. For instance, if the molding temperature in use was 280 °F, the levels in the experiment might be 275 °F and 285 °F, instead of the 260/300 °F contrast more typical of bold factor limits. Change levels in the other factors would be similarly small. The probability of creating a serious jump in scrap rate with any of the eight combinations in the design is then low, but so is the chance of being able to detect any factor effects from a single molding cycle at each combination.

However, the experiment is run on a much larger scale, at least a full shift of production for each of the eight treatments if not an entire day, or even a full week. When the sample size increases from 36 (one mold load) to somewhere in the neighborhood of 3,000 (one shift's output) or more, the percentage of scrap observed is almost certain to show whether any change in scrap rate has occurred, even if the change is not large.

When the eight extended runs are completed and the data can be analyzed, the significant effects will be detected. After the initial determination of which factors and interactions are important is made, planning out another run becomes much easier. If there had been no effect of time or holding pressure but the higher temperature and slower injection rate lowered scrap and also showed an interaction, the next experiment could be a full factorial on those two factors, with one level being the more productive of the two used in the first experiment and the second level being adjusted slightly in the direction that had showed improvement already. (For instance, temperature would be varied as 285/295 °F in the second-stage experiment.)

Use of a series of EVOP experiments can gradually take the process into its optimal region with very little negative impact on production. It does take more time than the standard DOX approach, but that may be an acceptable condition. Because the variation in the factor levels is not very large in EVOP practice, the changes in output will tend to be small. This means that the signal-to-noise ratio is not very favorable, and therefore the R-squared of the derived model is seldom very high, but, as discussed in Chapter 16, this does not mean the model is invalid, and the key quantities to be examined are the individual probability levels of the coefficients in the model. When they have high confidence levels, then the experiment has led to better understanding of the process, even if the total R-squared is not great.

The major pitfall of EVOP is that too many other things are allowed to change nonrandomly during experiments, such as batches of raw materials or environmental conditions that can confuse the data badly. This can be controlled with careful planning and the commitment of the production staff, so good communications and a team attitude are essential to an EVOP approach.

Finally, there occasionally arise situations that do not readily fit into any of the commonly used and programmed designs, such as when some or all of the different control factors cannot be used at the same number of levels. Designs then have to be customized for the situation, which is sometimes not too difficult and sometimes rather challenging.

One example was a situation in which one discrete factor had to be evaluated at six levels while two other continuous factors only needed to be screened at two levels. The full factorial for this would have taken 24 runs $(6 \times 2 \times 2)$, which would certainly have provided plentiful data on the process. However, with some thought it was possible to work up a half-factorial pattern that performed very satisfactorily with 50% of the effort. The actual pattern is shown in Fig. 22.1.

When the test data were available, they were easily interpreted through the use of basic graphs, such as the combination bar chart in Fig. 22.2 showing how the response of material dimensional stability was affected by all the factors at every level of each. It is immediately clear that continuous factor X has a substantial effect on the response, whereas factor Y has little, if any. Among the different discrete levels of factor Z, there is little question that D and E merit the most attention in whatever follow-up set of experiments would be drawn up next.

This kind of comparatively simple adaptation of designs is something that can be learned with time and experience, and some books are available with tables of mixed-level designs for reference. There are also whole families of designs that are less well known and used more seldom in industrial applica-

	Factor X	Factor Y	Factor Z (6 types)
Run 1	+	+	A
2	+	−	B
3	−	+	B
4	−	−	A
5	+	+	E
6	+	−	C
7	−	+	C
8	−	−	E
9	+	+	F
10	+	−	D
11	−	+	D
12	−	−	F

Figure 22.1 Condensed $6 \times 2 \times 2$ design of 12 runs

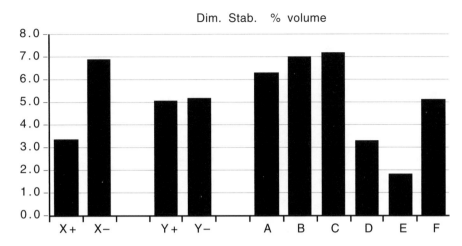

Figure 22.2 Combination bar chart displaying responses of material dimensional stability to all factors at every level

tions (such as balanced incomplete block designs); some of these may apply very well under particular circumstances, but they are not included in basic DOX training or most software programs because of their infrequent use.

In other, more complicated cases, it will turn out that drawing up the full pattern of some RSM design will call for some combinations of factor settings that may be very impractical for use or even impossible to run. This could be like trying to fill in the standard cube design only to find out that it is not possible to get data from one of the corners of the cube. In such situations, different strategies can be tried, such as running whatever point(s) are practical which are closest to the desired but unachievable set of factor settings. Some of the software packages used for RSM work will have built-in strategy for dealing with the problem.

If no apparently viable solution to special problems is readily found in the software manual or the literature, then it is time to either review and alter the experimental strategy or seek qualified expert help.

23 Computer Programs for Design of Experiments

The use of personal computers in design of experiments (DOX) has been a major factor in advancing the everyday use of the methodology, which for many years was largely limited to those well qualified in Statistics. Programs are available in varying degrees of sophistication and across a wide range of costs ($300–2,000). The bulk of the commercial software packages are PC based, with a few still running directly on a DOS system, although now the majority of current versions run through Windows, and there are at least two or three packages written for the Macintosh.

The very simplest way of using a computer to set up and analyze designed experiments is to conduct two-level arrays through a spreadsheet function. The patterns of the most frequently used two-level designs, including the 4-, 8-, 12-, and 16-run models, can be built into spreadsheet templates very easily. The template can then be altered to describe the particular experiment with the appropriate factor names and levels entered and a space supplied to enter results. This can then be printed out as a worksheet, with an assigned column for a randomized run order.

Once the work is done and the results have been recorded on the worksheet, they can be entered on the electronic template, which will automatically do the addition, subtraction, and division necessary to yield the effect levels for each factor. The templates can be used for full or half factorials or saturated fractional designs as well, as long as the columns are properly identified with their factors and interactions. The calculations are normally done using the factors at their coded levels, so the effects shown are normalized rather than relative to the units and levels of the control factors.

A scree chart can then be drawn up by hand or sometimes through the basic graphing functions built into most spreadsheets.

Naturally, only the simplest analysis of the data, which is rating the relative contributions of various factors and interactions, can be done this way. However, some commercial templates even have Taguchi subroutines built in so that not only major effects but also percent contribution and the specialized Taguchi signal-to-noise ratios are calculated. The simplest multilevel designs, namely the Taguchi L9 and L18 arrays, can also be set up in templates for a very elementary analysis.

Further, with the advent of more sophisticated spreadsheet programs, it is becoming possible to do analysis of variance (ANOVA) and regression analysis on experimental data from RSM designs as well. Very possibly, sets of DOX templates for major commercial spreadsheets will begin to take over part of the software market someday.

Specialized software packages for designed experiments normally contain three principal separate functions. The first is the collection of designs (or the algorithms for drawing up the designs) that the software writer has chosen as the ones around which the program will center. These are the patterns that can be very easily called up by the software user, like picking a particular screwdriver out of an available set of hand tools. Designs such as full and half factorials and Plackett-Burman designs are almost invariably included, going up to the size of 64-run experiments. Central composite and Box-Behnken patterns are also basic to any RSM subset, whereas face-centered cubic, Hartley, and d-optimal designs may or may not be included. They are usually set up for from three to seven control factors, sometimes for as many as 10.

Mixture designs are not always built in to the simpler programs, but every major RSM package will contain at least one or two forms of the simplex designs. Some have several variations available for the experimenter to choose from during the setup of the experiment. Combination designs that use both mixture and factorial aspects are not commonly seen but do exist in a few programs. It is also possible for an experienced user to adapt a factorial design to incorporate facets from a mixture design, even when the software is not set up for combination designs.

Taguchi's orthogonal arrays are most often found in software devoted to that methodology alone, but at least one general purpose package has a Taguchi module along with conventional factorial and simplex designs.

Some designs may not be found at all in software, such as the hyper-Graeco-Latin squares larger than the 3 × 3 (which is always in Taguchi systems as an L9). There is also a myriad of mixed-level designs in the literature, such as 16-run designs that contain one factor at eight levels and eight factors at two levels or two factors at four levels and nine factors at two levels, very few of which are ever built into software. (The Taguchi L18 is the only one regularly seen.)

The second major part of any DOX program is the analytical functions, which include a number of statistical subroutines such as linear regression, multifactor ANOVA, correlation matrices, data transformations, and a wide variety of assorted special correlations peculiar to the examination of test data and experiment characteristics. Some of these subroutines are quite simple and amount to no more than what was described above for a spreadsheet analysis of the data from a two-level experiment. Others can be quite complex, such as those that make use of fuzzy logic to hunt through a very large number of possible factor settings and their resulting effects on multiple response characteristics, each of which is related to the control factors by its own unique mathematical model. The goal of this exercise is to locate whatever combination of factor settings will produce the overall collection of responses that meet some restricted combination of desired responses. When the process is complex, numerous responses are involved, and the desired response combination is narrowly defined, this can amount to a level of simultaneous equation solving that would be utterly impractical for even a small team of mathematicians with calculators to perform in any reasonable time frame.

The software writer is compelled to build in all the major computing routines essential to data analysis from designed experiments, such as linear regression and ANOVA. Many of the lesser techniques are more or less optional, and no program will contain every subroutine known to statistical analysis. Some software packages are very generously equipped with extra techniques, far more than the average user will ever learn to utilize, but then the users at least have the option of exploring the variety of tools available to find those which fit their needs or preferences especially well. However, to make the best use of these assorted analytical tools, it usually becomes necessary to gain some more detailed knowledge of the statistical principles on which they are based.

This is where some software packages are less complete than they might be in that their manuals assume a level of statistical expertise on the part of the user that is beyond the majority of scientists and engineers. There is great value in having a readable manual, an excellent on-line help function, and good examples or a tutorial built into the program. Some programs are very usable only to those already very familiar with DOX, whereas others have been made so user friendly that not much more is needed than a knowledge of the process and of what constitutes good experimental practice in general for the user to select, run, and analyze the data of a perfectly valid designed experiment.

It is important to point out that the analytical functions are independent of the experimental designs built into the particular piece of DOX software. Experiments can be altered in a variety of ways or made into altogether new patterns, and routines such as regression or ANOVA will still work fine on the resulting set of test data. This means that if some program contained the 8- and 16-run Plackett-Burman designs but not the 12- or 20-run designs, with some effort the user could alter an existing pattern into one of the missing ones and use it accordingly. In most cases, the file from that altered experiment could then be used as a template for that design again if the need arose. Any central composite design could be made into the matching face-centered cubic pattern very easily, or into a Hartley design if that were desired.

This capacity to expand a program is not always recognized or used but can be worthwhile under some circumstances. For instance, few programs have mixed-level designs built in, but suppose someone needed to evaluate six factors, three of which were continuous and to be evaluated at three levels but the other three of which were discrete and only existing at two levels. (For instance, two catalyst types and the presence or absence of a final oven drying step.) The full-factorial experiment would call for 216 runs, which is $3 \times 3 \times 3 \times 2 \times 2 \times 2$.

Software could be used to set up a central composite design for the three continuous factors, which the experimenter would reduce to the Hartley pattern by editing the design. That would contain 14 different treatments (which includes a center point and three replicates). This pattern would be run at each of the four combinations of a half-factorial pattern of the three discrete factors, and all the data from the total of 56 runs would be entered into a single experiment file. This would contain all information needed to allow evaluation of all the factor contributions, two-factor interactions, and nonlinear terms, plus ample information for calculation of lack of fit and pure error, all of which the software would calculate validly for the experimenter.

The third part of most DOX programs is the graphics subsection, which makes it possible to plot graphs of varying types and complexity, now usually in colors, with customization of the axis scales, fonts, symbols, etc., possible as well. Use of these powerful graphing tools can make it easier for the experimenter to interpret the data and is especially useful in making professional presentations for management, customers, and other technical contributors. Mastery of this part of the software is not strictly necessary to good use of DOX (after all, designed experiments were developed and used effectively long before color pseudo-three-dimensional graphs of response surfaces were available), but it can be a handy tool.

Although the first and second parts of the program are of much more importance to its value as a real working tool, graphics capacity may be of significant worth when it comes to making presentations on some work to management or nontechnical personnel. An illustration of an intricate response surface will usually go a lot further than tables of data in helping people appreciate the complexity of a process and understand the scope of the work that has been done.

One especially helpful extra feature of a few programs is an optimization function. This comes into play after the experiment has been run and the data have supported the construction of response equations for several different characteristics of the process output. (That could mean several physical properties such as tensile strength, modulus, yield point, flex life, hardness, etc., other properties such as process yield, purity, throughput rate, and so on, or a combination of physical and process properties.)

An optimization routine helps the experimenter balance all the desired properties in the overall best combination possible. When there are several targeted properties, some of which may work against each other, finding the optimal combination can be a major challenge. In such cases, the value of an optimization function becomes considerable.

There are at least two dozen different DOX software programs available at this time, some from well-known firms associated with statistical programs, some from large companies that developed the software originally for in-house use but now hold training seminars and sell the package as part of the course, and some from various smaller enterprises selling directly or through publishers and distributors. DOX software applications vary from very minimal packages of templates for use with commercial spreadsheets, to larger programs with a good range of designs and diagnostic tools, to very extensive general purpose statistical programs with more than enough features to please a professional statistician with a concentration in DOX. The large statistical programs that have DOX modules or subsections built in tend to be more expensive than the programs that are tailored very specifically to DOX, but for those who will make use of the other capabilities of such expanded applications, the investment may be well worthwhile. On the other hand, the specific DOX programs tend to be more user friendly than the full-service statistical programs, which can be of considerable importance to the neophyte user.

There is not a good correlation between the cost of DOX software and its value to the buyer. Depending on who is to use the program and in what setting, a $400 investment in a reasonably basic package might be much more cost effective than $1,500 for a deluxe program. The features that need to be evaluated in relation to the company doing the purchasing and its people and their intended applications of DOX are (roughly in order of importance):

- Degree of user friendliness;
- usability and completeness of the manual;

- on-line help function and tutorial;
- types of designs included;
- types of analysis (especially residuals) and their ease of use;
- availability of technical support;
- special features (optimization subroutine, mixed designs, etc.);
- graphics capability; and
- cost (including network version if needed).

A few of the major programs now on the market in the United States are described very briefly below:

- Design Expert (v. 5): DOX only, very user friendly, numerous design options, readily customizable for special designs, multiple response optimization capability.
- E-Chip: DOX only, less user friendly, numerous designs including mixture/factorial combinations and Taguchi, but not customizable, multiple response optimization capability.
- JMP: statistical program, moderately user friendly, numerous design options, optimization capability
- Minitab: statistical program, very user friendly, numerous design options, graphic optimization
- Statgraphics: statistical program, very user friendly, numerous design options, graphic optimization
- Statistica: statistical program, moderately user friendly, numerous design options, graphic optimization

How well people will make use of any piece of DOX software depends on their need and/or motivation, its ease of use and/or good training in its use, and a very clear understanding of the underlying principles that must be applied in setting the experimental parameters and performing the actual experiments and testing. The best software in the existence is still totally subject to the GIGO syndrome (garbage in, garbage out). The use of a flawed experimental strategy, an incomplete set of control factors, inappropriate factor levels, sloppy experimental technique, or insensitive test methods will inevitably impair or destroy the value of any data generated, and no analytical subroutine can really make up for low-quality data.

There is no doubt that use of good DOX software is enormously valuable in exploring processes of moderate to high complexity. As with any acquisition of significant equipment or tools, the buyers are well advised to consider first their immediate and future needs and then match those needs up carefully to products on the market.

References

1. Barker, Thomas B. (1985) *Quality By Experimental Design*. New York and Basel: Marcel Dekker
2. Bhote, Keki R. (1991) *World Class Quality*. New York: ASQC Quality Press
3. Box, G.E.P., Hunter, G.S., and Hunter, W.G. (1978) *Statistics for Experimenters*. New York: Wiley
4. Chalmer, Bruce J. (1987) *Understanding Statistics*. New York: Marcel Dekker
5. Cheremisinoff, Nicholas P. (1987) *Practical Statistics for Engineers and Scientists*. Lancaster, PA: Technomic
6. Cochrane, W. G., and Cox, G. M. (1957) *Experimental Designs*. New York: Wiley
7. Cornell, John A. (1990) *Experiments with Mixtures*. New York: Wiley
8. Lochner, R. H., and Matar, J. E. (1990) *Designing for Quality*. New York: ASQC Quality Press
9. Ott, Lyman. (1988) *An Introduction to Statistical Methods and Data Analysis*. Boston: PWS-Kent
10. Petersen, Roger G. (1985) *Design and Analysis of Experiments*. New York and Basel: Marcel Dekker
11. Ross, Philip J. (1988) *Taguchi Techniques for Quality Engineering*. New York: McGraw-Hill

Glossary

Analysis of variance (ANOVA)   A statistical technique for comparing a number of groups to determine whether they all may be part of a single parent population

Bias error   Systematic variation in process output caused by an unintended pattern of changes in process conditions

Blocking   Deliberate association of some elements of an experimental design so that a particular contrast is reinforced in the design (four different kinds of horseshoes tested by putting one each on the feet of every test horse) or an unavoidable contrast is specifically taken into account and quantified (assigning test day as a variable in the design when it is necessary to do two days of testing instead of one for some set of experimental parts)

Box-Behnken design   An advanced design that uses three levels of each factor to generate a full date set suitable for response surface methods

Central composite design   An advanced design that uses five levels of each factor to generate a full date set suitable for response surface methods

Central Limit Theorem   A mathematical principle stating that averages are much more reliable indicators of the value of some property than any individual measurement, and the larger the sample size used to calculate the average, the better the estimate it gives

Coding   Use of special designations for variable levels to make examination of the pattern easier, e.g., -1 and $+1$ instead of 350 °F and 400 °F in a design

Confidence interval   A range of values within which a particular number of interest is calculated to fall, at some specific level of probability such as 90% or 95%

Confounding   A lack of capacity for separately analyzing effects of factors and/or interactions caused by the use of an experimental pattern that has those effects located so that the mathematical techniques for evaluating one are exactly the same as for the other(s); also known as aliasing

Continuous factor   A control variable that can be changed over many levels, such as voltage, temperature, or percent of some ingredient

Degrees of freedom (df)   A measure of the amount of information available in some data set; most of the time df is equal to the number of experiments minus 1, thus an 8-run design has 7 df, a 16-run has 15 df, etc.

Designed experiment   A single set of actual related experiments drawn up from one of the numerous types of designs to be found in the body of methods for design of experiments

Design of experiments (DOX, DOE)   A whole set of methods for doing experiments in patterns instead of one at a time, with the benefit being that much better and much more information can be gotten from the resulting pattern of test data than from more conventional kinds of experimenting

Discrete factor   A factor that only comes at clearly separate levels, such as different kinds of chemicals, different operators, etc.

Distribution   Pattern in which numbers in a group vary within the overall grouping, such as all numbers appearing equally (a block distribution) or most of the numbers appearing toward the middle of the group and fewer appearing toward each end of the grouping (a normal distribution)

Experimental error   This has nothing to do with making mistakes; it is simply the normal amount of variation seen in test results from experiments or measurements all done at the same conditions; also known as scatter or experimental noise

Experimental run   A particular set of conditions used in the process being investigated for which 2–32 individual different sets of conditions may be used to make up the overall experimental pattern; also called the treatment, experimental setup, etc.

F-test   A comparison of two sample group variances to determine whether they are close enough in size to have come from the same parent population

Face-centered cubic design   An advanced design that uses three levels of each factor to generate a full data set suitable for response surface methods

Factor   A process condition that significantly affects or controls the process output, such as temperature, pressure, type of raw material, concentration of active ingredient, etc.; formally called an independent variable, control variable, etc.

Fractional-factorial design   A partial set of experiments, still fully balanced in terms of appearance of factors and levels, based on but not as large as a full factorial design

Full-factorial design   A set of experiments that includes every level of every factor with every other factor and level combination

Half-factorial design   The most basic fractional-factorial design, in which the number of experimental runs is one-half that of a full factorial (only simple confounding patterns are found)

Histogram   A vertical bar graph in which the width of the bars represents a subinterval of possible values from a larger group and the height of each bar is proportional to the number of values in the group which fall into the particular subinterval assigned to that bar; the most common method of graphically representing a distribution

Hyper-Graeco-Latin squares   A series of multilevel screening designs that permit examination of N variables at $N - 1$ levels, with the most commonly known example being the Taguchi L9, which can have up to four variables each at three levels; these designs have no capacity at all for detecting interactions but can be quite useful for screening a series of discrete variables and/or wide ranges in levels of use of variables when very little is known of the process

Interaction   The working together of different control factors so that their total effect together is not just the sum of their separate effects but is instead much larger or smaller than would be expected

Mean   The average of any group of numbers, that is, the sum of all individuals in the group divided by the number of individuals in the group

Median   The center individual number in a group of numbers that have been arranged in order of size; if there are an even number of individuals, the median becomes the average of the two center individual numbers

Mixture designs   Families of designs for which the levels of the control factors are not independent, so all the various treatments used are directly related to each other in the way they combine the factor levels; typically, the factors are ingredients in a mixture that contains a total of 100% of the ingredients in combination and each factor level is the percentage of that factor in the particular experimental treatment

Mode   The number that appears most often in a larger group of numbers; if more than one individual number stands out clearly from the group, the distribution can be termed multimodal

Mu   The average of a full population, whether real or hypothetical; seldom if ever truly known but usually the object of various methods of estimation such as use of the average of some sample group, termed $\overline{X}$, to indicate approximately where mu may be

Normal distribution   A pattern of individual numbers within a large group such that the density of numbers increases rapidly from the smallest ones toward the average and then decreases just as rapidly from the average to the largest numbers; also called a Gaussian distribution after the mathematician who derived the equation that fits the ideal form of this distribution (nonnormal distributions include the binomial, Poisson, and Weibull types)

Parent population   The real or hypothetical group containing all possible members fitting the description of the group; seldom actually known but the focus of the bulk of statistical techniques

Plackett-Burman designs   A series of fractional factorial designs developed specifically for screening effects of main control factors; the series includes 8-, 12-, 16-, 20-, 24-, and 32-run patterns, with the 12-, 20-, and 24-run types being termed nongeometric, meaning all interaction effects are dispersed broadly throughout the data and therefore are less likely to interfere with their analysis

Random error   Differences in process output caused only by the normal amount of variation inherent to the basic process

Reflected design   A fractional-factorial design that contains the same factors, range of levels, and number of experimental runs as another design but has the levels of each factor reversed in every run; if in the original design the second run was set up with the factors at their +1, −1, +1, and −1 levels, in the reflected design they would be −1, +1, −1, and +1 (The two designs are called complementary and are reverse images of each other.)

Replication   Duplication of a previously run treatment, usually performed to gain some measure of the amount of variation natural to the process over time; as contrasted to repetition, which is the testing of more than one item of output from a single run of the process, which measures the combination of test error and short-term process variation

Response   A characteristic or property of the process output which is of interest and changes in which are used to analyze how the process really functions; formally termed a dependent variable, the experimental signal, a characteristic, a result, etc.

Response Surface Methodology (RSM)   Techniques for mapping the multidimensional pattern of responses to varying levels of several control factors; RSM is dependent on the use of regression analysis on data from experiments at three or more levels and can be used to find minima or maxima in the response patterns so as to allow optimization of the process

Robust process   A term popularized by Dr. Taguchi referring to a process that has been made relatively insensitive to minor changes in some factor levels by selection of particular levels of other factors that affect process variability more than process average output

Sampling distribution   Distribution of all possible samples of a certain size that could be taken from some parent population; because there are more possible combinations of samples than of individuals in the population, the sampling distribution is larger and more narrowly distributed than the parent population (which is the basis for the Central Limit Theorem)

Saturated design   A design in which every column for factor settings has been assigned to an individual factor with no columns left open for interaction or error terms

Scree plot   A plot of relative factor effect levels graded in decreasing order, used to separate out real from minor or scatter effects

Screening design   Any fractional-factorial experimental design being used solely or primarily to allow evaluation of the main effects of several control factors, with full awareness that effects of any possible interactions between factors are being confounded with main factor effects; these are useful in initial evaluations of a relatively unknown process or set of control factors (see Plackett-Burman and Taguchi designs)

Sigma   Greek term used to denote the standard deviation of a full population, real or theoretical; often used inaccurately to indicated what is really a sample standard deviation, which should be referred to as s

Signal-to-noise ratio   The relative size of an effect as compared to background scatter in the response being measured

Statistics   A subsection of the science of mathematics concerned with the collection, organization, analysis, interpretation, and presentation of data

Statistic   Any number measuring something, such as an animal's weight, an athlete's batting average, a car's gas mileage rating, etc.

Standard deviation (Std. Dev., SD)   A measure of the width of a normal distribution, derived as the square root of the variance of that distribution; normal distributions are approximately six standard deviations in width, because ideally that overall width contains 99.7% of the numbers in the distribution

Standard error   The theoretical standard deviation of a sampling distribution, usually estimated with the use of a sample standard deviation and the square root of the size of the sample group; useful in calculating confidence intervals

t-test   Use of revised estimates of distribution width to generate more accurate confidence intervals from limited sample groups; allows better comparisons of sample means to determine whether the different samples are likely to fall within the same parent population

Taguchi designs   A series of designs all related to Plackett-Burman or other classic screening designs which uses different conventions and slightly different analytical techniques; when used in processes in which interactions are likely to exist, the patterns have to be expanded until they become very similar to other classic patterns; other parts of the Taguchi techniques (which make up a sizable body of philosophical principles along with patterns and analysis) involve more specific examination of signal-to-noise considerations and process optimization using concepts of robust conditions

Unsaturated design   A fractional-factorial design in which at least some columns remain unassigned to individual control factors, which allows some estimates to be made on interactions or scatter levels in the experiment

Variance   The mathematically preferred estimate of the width of a distribution, derived by subtracting the average of the distribution from each individual member, squaring the resulting numbers, adding up all the squares, and then dividing by the number of all the individuals (except when any sort of sample group is being examined instead of the entire population; for sample groups the number of individuals minus one is used for the division)

Appendix

Several standardized tables of experimental designs are included for reference on the following pages.

Factor	A	B	C	D	E	F	G	H	I	J	K	L	M	N	O
4-run															
Run #1	+	+	+												
2	+	−	−												
3	−	+	−												
4	−	−	+												
8-run															
Run #1	+	+	+	+	+	+	+								
2	+	+	−	+	−	−	−								
3	+	−	+	−	+	−	−								
4	+	−	−	−	−	+	+								
5	−	+	+	−	−	+	−								
6	−	+	−	−	+	−	+								
7	−	−	+	+	−	−	+								
8	−	−	−	+	+	+	−								
9-run (full factorial)															
Run #1	−	−													
2	−	0													
3	−	+													
4	0	−													
5	0	0													
6	0	+													
7	+	−													
8	+	0													
9	+	+													

9-run (Taguchi L9)

Factor	A	B	C	D	E	F	G	H	I	J	K	L	M	N	O
Run #1	-	-	-	-											
2	-	0	0	0											
3	-	+	+	+											
4	0	-	0	+											
5	0	0	+	-											
6	0	+	-	0											
7	+	-	+	0											
8	+	0	-	+											
9	+	+	0	-											

12-run (Plackett-Burman nongeometric screening design)

Factor	A	B	C	D	E	F	G	H	I	J	K	L	M	N	O
Run #1	+	+	-	+	+	+	-	-	-	+	-				
2	+	-	+	+	+	-	-	-	+	-	+				
3	-	+	+	+	-	-	-	+	-	+	+				
4	+	+	+	-	-	-	+	-	+	+	-				
5	+	+	-	-	-	+	-	+	+	-	+				
6	+	-	-	-	+	-	+	+	-	+	+				
7	-	-	-	+	-	+	+	-	+	+	+				
8	-	-	+	-	+	+	-	+	+	+	-				
9	-	+	-	+	+	-	+	+	+	-	-				
10	+	-	+	+	-	+	+	+	-	-	-				
11	-	+	+	-	+	+	+	-	-	-	+				
12	-	-	-	-	-	-	-	-	-	-	-				

Factor	A	B	C	D	E	F	G	H	I	J	K	L	M	N	O
16-run Run #1	+	+	+	+	+	+	+	+	+	+	+	+	+	+	+
2	+	+	+	−	+	+	−	+	−	−	+	−	−	−	−
3	+	+	−	+	+	−	+	−	+	−	−	+	−	−	−
4	+	+	−	−	−	−	−	−	−	+	−	−	+	+	+
5	+	−	+	+	−	+	+	−	−	+	−	−	+	−	−
6	+	−	+	−	−	+	−	−	+	−	−	+	−	+	+
7	+	−	−	+	−	−	+	+	−	−	+	−	−	+	+
8	+	−	−	−	−	−	−	+	+	+	+	+	+	−	−
9	−	+	+	+	−	−	−	+	+	+	−	−	−	+	−
10	−	+	+	−	−	−	+	+	−	−	−	+	+	−	+
11	−	+	−	+	−	−	−	−	+	−	+	−	+	−	+
12	−	+	−	−	−	+	+	−	−	+	+	+	−	+	−
13	−	−	+	+	+	+	−	−	−	+	+	+	−	−	+
14	−	−	+	−	+	−	+	−	+	−	+	−	+	+	−
15	−	−	−	+	+	−	−	+	−	−	−	+	+	+	−
16	−	−	−	−	+	+	+	+	+	+	−	−	−	−	+

18 run (Taguchi multi-level screening design)

Factor	A	B	C	D	E	F	G	H	I	J	K	L	M	N	O
Run #1	-	-	-	-	-	-	-	-							
2	-	-	O	O	O	O	O	O							
3	-	-	+	+	+	+	+	+							
4	-	O	-	-	O	O	+	+							
5	-	O	O	O	+	+	-	-							
6	-	O	+	+	-	-	O	O							
7	-	+	-	O	-	+	O	+							
8	-	+	O	+	O	-	+	-							
9	-	+	+	-	+	O	-	O							
10	+	-	-	+	+	O	O	-							
11	+	-	O	-	-	+	+	O							
12	+	-	+	O	O	-	-	+							
13	+	O	-	O	+	-	+	O							
14	+	O	O	+	-	O	-	+							
15	+	O	+	-	O	+	O	-							
16	+	+	-	+	O	+	-	O							
17	+	+	O	-	+	-	O	+							
18	+	+	+	O	-	O	+	-							

20-run (Plackett-Burman nongeometric screening design)

Factor Run #1	A	B	C	D	E	F	G	H	I	J	K	L	M	N	O	P	Q	R	S
1	+	+	−	−	+	+	+	+	−	+	−	+	−	−	−	−	+	+	−
2	+	−	−	+	+	+	+	−	+	−	+	−	−	−	−	+	+	−	+
3	−	−	+	+	+	+	−	+	−	+	−	−	−	−	+	+	−	+	+
4	−	+	+	+	+	−	+	−	+	−	−	−	−	+	+	−	+	+	−
5	+	+	+	+	−	+	−	+	−	−	−	−	+	+	−	+	+	−	−
6	+	+	+	−	+	−	+	−	−	−	−	+	+	−	+	+	−	−	+
7	+	+	−	+	−	+	−	−	−	−	+	+	−	+	+	−	−	+	+
8	+	−	+	−	+	−	−	−	−	+	+	−	+	+	−	−	+	+	+
9	−	+	−	+	−	−	−	−	+	+	−	+	+	−	−	+	+	+	+
10	+	−	+	−	−	−	−	+	+	−	+	+	−	−	+	+	+	+	−
11	−	+	−	−	−	−	+	+	−	+	+	−	−	+	+	+	+	−	+
12	+	−	−	−	−	+	+	−	+	+	−	−	+	+	+	+	−	+	−
13	−	−	−	−	+	+	−	+	+	−	−	+	+	+	+	−	+	−	+
14	−	−	−	+	+	−	+	+	−	−	+	+	+	+	−	+	−	+	−
15	−	−	+	+	−	+	+	−	−	+	+	+	+	−	+	−	+	−	−
16	−	+	+	−	+	+	−	−	+	+	+	+	−	+	−	+	−	−	−
17	+	+	−	+	+	−	−	+	+	+	+	−	+	−	+	−	−	−	−
18	+	−	+	+	−	−	+	+	+	+	−	+	−	+	−	−	−	−	+
19	−	+	+	−	−	+	+	+	+	−	+	−	+	−	−	−	−	+	+
20	−	−	−	−	−	−	−	−	−	−	−	−	−	−	−	−	−	−	−

Basic Response Surface Designs

Face-Centered Cubic Design

Run #	A	B	C
1	-1	-1	-1
2	1	-1	-1
3	-1	1	-1
4	1	1	-1
5	-1	-1	1
6	1	-1	1
7	-1	1	1
8	1	1	1
9	-1	0	0
10	1	0	0
11	0	-1	0
12	0	1	0
13	0	0	-1
14	0	0	1
15	0	0	0
16	0	0	0
17	0	0	0
18	0	0	0
19	0	0	0
20	0	0	0

Central Composite Design

Run #	A	B	C
1	-1	-1	-1
2	1	-1	-1
3	-1	1	-1
4	1	1	-1
5	-1	-1	1
6	1	-1	1
7	-1	1	1
8	1	1	1
9	-1.682	0	0
10	1.682	0	0
11	0	-1.682	0
12	0	1.682	0
13	0	0	-1.682
14	0	0	1.682
15	0	0	0
16	0	0	0
17	0	0	0
18	0	0	0
19	0	0	0
20	0	0	0

Box-Behnken Design

Run #	A	B	C
1	1	1	0
2	1	-1	0
3	-1	1	0
4	-1	-1	0
5	1	0	-1
6	1	0	1
7	-1	0	-1
8	-1	0	1
9	0	1	-1
10	0	1	1
11	0	-1	-1
12	0	-1	1
13	0	0	0
14	0	0	0
15	0	0	0

Face-Centered Cubic Design

Run #	A	B	C	D
1	-1	-1	-1	-1
2	1	-1	-1	-1
3	-1	1	-1	-1
4	1	1	-1	-1
5	-1	-1	1	-1
6	1	-1	1	-1
7	-1	1	1	-1
8	1	1	1	-1
9	-1	-1	-1	1
10	1	-1	-1	1
11	-1	1	-1	1
12	1	1	-1	1
13	-1	-1	1	1
14	1	-1	1	1
15	-1	1	1	1
16	1	1	1	1
17	-1	0	0	0
18	1	0	0	0
19	0	-1	0	0
20	0	1	0	0
21	0	0	-1	0
22	0	0	1	0
23	0	0	0	-1
24	0	0	0	1
25	0	0	0	0
26	0	0	0	0
27	0	0	0	0
28	0	0	0	0
29	0	0	0	0
30	0	0	0	0
31	0	0	0	0

Central Composite Design

Run #	A	B	C	D
1	-1	-1	-1	-1
2	1	-1	-1	-1
3	-1	1	-1	-1
4	1	1	-1	-1
5	-1	-1	1	-1
6	1	-1	1	-1
7	-1	1	1	-1
8	1	1	1	-1
9	-1	-1	-1	1
10	1	-1	-1	1
11	-1	1	-1	1
12	1	1	-1	1
13	-1	-1	1	1
14	1	-1	1	1
15	-1	1	1	1
16	1	1	1	1
17	-2	0	0	0
18	2	0	0	0
19	0	-2	0	0
20	0	2	0	0
21	0	0	-2	0
22	0	0	2	0
23	0	0	0	-2
24	0	0	0	2
25	0	0	0	0
26	0	0	0	0
27	0	0	0	0
28	0	0	0	0
29	0	0	0	0
30	0	0	0	0
31	0	0	0	0

Box-Behnken Design

Run #	A	B	C	D
1	1	1	0	0
2	1	-1	0	0
3	-1	1	0	0
4	-1	-1	0	0
5	0	0	1	1
6	0	0	1	-1
7	0	0	-1	1
8	0	0	-1	-1
9	0	0	0	0
10	1	0	0	1
11	1	0	-1	-1
12	1	0	1	-1
13	0	1	1	0
14	0	1	-1	0
15	0	-1	1	0
16	0	-1	-1	0
17	1	0	1	0
18	1	0	-1	0
19	-1	0	1	0
20	-1	0	-1	0
21	0	1	0	1
22	0	1	0	-1
23	0	-1	0	1
24	0	-1	0	-1
25	0	0	0	0
26	0	0	0	0
27	0	0	0	0

Face-Centered Cubic Design

Run #	A	B	C	D	E
1	-1	-1	-1	-1	1
2	1	-1	-1	-1	-1
3	-1	1	-1	-1	-1
4	1	1	-1	-1	1
5	-1	-1	1	-1	-1
6	1	-1	1	-1	1
7	-1	1	1	-1	1
8	1	1	1	-1	-1
9	-1	-1	-1	1	-1
10	1	-1	-1	1	1
11	-1	1	-1	1	1
12	1	1	-1	1	-1
13	-1	-1	1	1	1
14	1	-1	1	1	-1
15	-1	1	1	1	-1
16	1	1	1	1	1
17	-1	0	0	0	0
18	1	0	0	0	0
19	0	-1	0	0	0
20	0	1	0	0	0
21	0	0	-1	0	0
22	0	0	1	0	0
23	0	0	0	-1	0
24	0	0	0	1	0
25	0	0	0	0	-1
26	0	0	0	0	1
27	0	0	0	0	0
28	0	0	0	0	0
29	0	0	0	0	0
30	0	0	0	0	0
31	0	0	0	0	0
32	0	0	0	0	0

Central Composite Design

Run #	A	B	C	D	E
1	-1	-1	-1	-1	1
2	1	-1	-1	-1	-1
3	-1	1	-1	-1	-1
4	1	1	-1	-1	1
5	-1	-1	1	-1	-1
6	1	-1	1	-1	1
7	-1	1	1	-1	1
8	1	1	1	-1	-1
9	-1	-1	-1	1	-1
10	1	-1	-1	1	1
11	-1	1	-1	1	1
12	1	1	-1	1	-1
13	-1	-1	1	1	1
14	1	-1	1	1	-1
15	-1	1	1	1	-1
16	1	1	1	1	1
17	-2	0	0	0	0
18	2	0	0	0	0
19	0	-2	0	0	0
20	0	2	0	0	0
21	0	0	-2	0	0
22	0	0	2	0	0
23	0	0	0	-2	0
24	0	0	0	2	0
25	0	0	0	0	-2
26	0	0	0	0	2
27	0	0	0	0	0
28	0	0	0	0	0
29	0	0	0	0	0
30	0	0	0	0	0
31	0	0	0	0	0
32	0	0	0	0	0

Box-Behnken Design					
Run #	A	B	C	D	E
1	1	1	0	0	0
2	1	-1	0	0	0
3	-1	1	0	0	0
4	-1	-1	0	0	0
5	0	0	1	1	0
6	0	0	1	-1	0
7	0	0	-1	1	0
8	0	0	-1	-1	0
9	0	1	0	0	1
10	0	1	0	0	-1
11	0	-1	0	0	1
12	0	-1	0	0	-1
13	1	0	1	0	0
14	1	0	-1	0	0
15	-1	0	1	0	0
16	-1	0	-1	0	0
17	0	0	0	1	1
18	0	0	0	1	-1
19	0	0	0	-1	1
20	0	0	0	-1	-1
21	0	0	0	0	0
22	0	0	0	0	0
23	0	0	0	0	0
24	0	1	1	0	0
25	0	1	-1	0	0
26	0	-1	1	0	0
27	0	-1	-1	0	0
28	1	0	0	1	0
29	1	0	0	-1	0
30	-1	0	0	1	0
31	-1	0	0	-1	0
32	0	0	1	0	1
33	0	0	1	0	-1
34	0	0	-1	0	1
35	0	0	-1	0	-1
36	1	0	0	0	1
37	1	0	0	0	-1
38	-1	0	0	0	1
39	-1	0	0	0	-1
40	0	1	0	1	0
41	0	1	0	-1	0
42	0	-1	0	1	0
43	0	-1	0	-1	0
44	0	0	0	0	0
45	0	0	0	0	0
46	0	0	0	0	0

Index